Advances in Ocean Mapping and Nautical Cartography

Advances in Ocean Mapping and Nautical Cartography

Guest Editors

Giuseppe Masetti
Ian Church
Anand Hiroji
Ove Andersen

MDPI

Basel • Beijing • Wuhan • Barcelona • Belgrade • Novi Sad • Cluj • Manchester

Guest Editors
Giuseppe Masetti
Center for Coastal and Ocean Mapping and NOAA-UNH Joint Hydrographic Center
University of New Hampshire
Durham
USA

Ian Church
Department of Geodesy and Geomatics Engineering
University of New Brunswick
Fredericton
Canada

Anand Hiroji
School of Ocean Science and Engineering
University of Southern Mississippi
Hattiesburg
USA

Ove Andersen
Department of Computer Science
Aalborg University
Aalborg East
Denmark

Editorial Office
MDPI AG
Grosspeteranlage 5
4052 Basel, Switzerland

This is a reprint of the Special Issue, published open access by the journal *Geomatics* (ISSN 2673-7418), freely accessible at: https://www.mdpi.com/journal/geomatics/special_issues/ocean_mapping.

For citation purposes, cite each article independently as indicated on the article page online and as indicated below:

Lastname, A.A.; Lastname, B.B. Article Title. *Journal Name* **Year**, *Volume Number*, Page Range.

ISBN 978-3-7258-3069-5 (Hbk)
ISBN 978-3-7258-3070-1 (PDF)
https://doi.org/10.3390/books978-3-7258-3070-1

Cover image courtesy of Giuseppe Masetti

Contents

Preface

This Special Issue showcases cutting-edge research and technological innovations enhancing our understanding of ocean mapping. Furthermore, the collection emphasizes novel methodologies and tools, which are critical for sustainable ocean management, safety in maritime navigation, and environmental protection.

With contributions spanning bathymetric surveys, geospatial data integration, and advancements in remote sensing, these articles demonstrate significant strides made in the fields of ocean mapping and related hydrospatial applications. This Special Issue – dedicated to researchers, industry professionals, and policymakers – will advance knowledge on ocean mapping and enhance the management of marine environments.

Giuseppe Masetti, Ian Church, Anand Hiroji, and Ove Andersen
Guest Editors

geomatics

MDPI

Editorial

Advancements in Ocean Mapping and Nautical Cartography

Giuseppe Masetti [1,2,*], Ian Church [3], Anand Hiroji [4] and Ove Andersen [2,5]

1 Center for Coastal and Ocean Mapping and NOAA-UNH Joint Hydrographic Center, University of New Hampshire, Durham, NH 03824, USA
2 Danish Hydrographic Office, Danish Geodata Agency, 9400 Nørresundby, Denmark; ovand@gst.dk
3 Department of Geodesy and Geomatics Engineering, University of New Brunswick, Fredericton, NB E3B5A3, Canada; ian.church@unb.ca
4 School of Ocean Science and Engineering, University of Southern Mississippi, Stennis Space Center, Hattiesburg, MS 39406, USA; anand.hiroji@usm.edu
5 Department of Computer Science, Aalborg University, Selma Lagerlöfs Vej 300, 9220 Aalborg East, Denmark
* Correspondence: gmasetti@ccom.unh.edu

Citation: Masetti, G.; Church, I.; Hiroji, A.; Andersen, O. Advancements in Ocean Mapping and Nautical Cartography. *Geomatics* **2024**, *4*, 433–436. https://doi.org/10.3390/geomatics4040023

Received: 1 November 2024
Accepted: 26 November 2024
Published: 28 November 2024

1. Introduction

Ocean mapping and nautical cartography are foundational to understanding and managing marine environments. These fields support a wide range of applications—from safe navigation and resource management to ecosystem conservation and assessment of the impacts of climate change [1]. The ocean remains largely uncharted at a high resolution, and new technologies are crucial to bridge this knowledge gap [2]. This Special Issue of *Geomatics* brings together nine studies, each tackling unique challenges facing ocean mapping. The articles highlight a wide-ranging spectrum of innovations, from automated data processing and high-resolution mapping systems to ecosystem-based planning and machine learning applications for habitat classification. Together, they showcase how diverse, interdisciplinary approaches are essential to advancing oceanography and marine spatial planning, facilitating sustainable development, and contributing to safer and more informed approaches to ocean use.

The collected works underscore the diversity of current research into ocean mapping and cartography, with methodologies ranging from traditional sonar-based mapping to advanced, machine learning-driven analyses of marine habitats. By exploring different technological and methodological advancements, this Special Issue emphasizes the importance of both preserving historical data and embracing new tools and approaches to meet the evolving demands of marine science and resource management [3]. This Editorial review provides an overview of these articles, highlighting their collective contribution to the field and their implications for the future of ocean mapping.

2. Data Collection and Review

2.1. A New Approach to Wide-Area Deep-Ocean Mapping

In response to the challenges of mapping deep-sea areas from the ocean surface, Ryu et al. [4] present a cutting-edge system that employs autonomous surface vessels to deploy a distributed array of sonar sensors. This design overcomes the limitations of sonar-based underwater mapping by achieving a high resolution at significant depths without the need to deploy costly and logistically complex underwater vessels. The sparse-array approach allows for the precise tracking of sonar positions relative to each other, compensating for oceanic conditions, and thereby advancing bathymetric mapping from the surface to unprecedented depths.

2.2. Open-Source Tools for Sonar Acceptance Testing

In the field of sonar data acquisition, testing and troubleshooting are critical for ensuring data accuracy. Younkin and Umfress [5] showcase the usefulness of Kluster, an open-source Python-based software, which streamlines sonar data processing and visualization, particularly for the Kongsberg EM712 system. By offering a user-friendly interface that supports common sonar tests, Kluster enables field professionals and scientists alike to easily evaluate and verify sonar system data, making it a valuable asset for marine data collection operations.

2.3. *Automating Hydrographic Data Review for Nautical Cartography*

Hydrographic data reviews involve a meticulous process that includes human error risks and substantial time investments. Masetti et al. [6] address these issues by introducing automated review tools that incorporate NOAA's hydrographic specifications, providing a standardized and efficient solution for hydrographic offices. These tools—named QC Tools and CA Tools—accelerate the data review process and improve reproducibility, serving as a model for conducting automated data reviews across the hydrographic community. This development marks a significant step forward in processing efficiency and data quality, improving navigation safety.

3. Bathymetric Modeling: Data Fusion and Interpolation

3.1. *Automating Historical Data Compilation with Shom's Téthys Workflow*

The French Hydrographic Service, Shom, has compiled over 300 years of data into an automated workflow known as Téthys. Le Deunf et al. [7] detail how this workflow integrates historical data from multiple sources, enhancing the accuracy of nautical charts by consolidating bathymetric data. The project emphasizes the importance of data fusion and ensures that modern digital models leverage centuries of seafloor data for current applications, demonstrating the importance of historical continuity in hydrographic sciences.

3.2. *Multigrid/Multiresolution Interpolation to Reduce Data Artifacts*

Traditional interpolation techniques often create artifacts such as over-smoothing in regions with a high data density. Rodriguez-Perez and Sanchez-Carnero [8] tackle this with a new multigrid interpolation approach that dynamically adapts to the data density, ensuring detail preservation in data-dense regions and achieving a smoother surface where data are sparse. This methodology allows for the creation of accurate digital elevation models (DEMs), improving seafloor mapping accuracy and providing realistic error estimations for applications like hazard mapping and habitat analysis.

3.3. *Denmark's Digital Bathymetric Model*

In a comprehensive mapping of Danish waters, Masetti et al. [9] compile a 50 m resolution bathymetric model using datasets spanning decades. This model serves multiple purposes, from environmental management to infrastructure development, highlighting the value of accessible, high-resolution bathymetric models in marine policy and planning. This digital bathymetric model (DBM) also offers web access through the Danish Geodata Agency, illustrating how data transparency and accessibility are integral to effective spatial planning. Notably, the insights and methodologies presented in this work provided a foundational basis for the authors' subsequent article detailing the release of an enhanced version the model, further refining its applications in marine and spatial planning [10].

3.4. *Mapping the Kerguelen Plateau's Tectonic and Bathymetric Structures*

In an exploration of one of the world's most remote plateaus, Lemenkova [11] integrates satellite and marine geophysical datasets to map the complex structures within the Kerguelen Plateau. This study sheds light on the plateau's tectonic history and heterogeneous seafloor composition, enriching our scientific understanding of marine geology. By overlaying magnetic, gravitational, and topographic data, this study exemplifies how advanced cartographic techniques can elucidate the geological development of underwater landforms.

4. Spatial Planning and Machine Learning for Habitat Classification

4.1. Smart Marine Ecosystem-Based Planning (SMEP) for Greece

Recognizing the importance of data-driven marine governance, Contarinis et al. [12] introduce the SMEP framework, a spatial planning approach tailored to Greek marine ecosystems. The model emphasizes the importance of iterative planning cycles and continuous environmental monitoring to ensure sustainable development. SMEP brings together ecological data and human activity metrics, fostering a comprehensive planning process that reflects Greece's unique coastal environment. This approach aims to serve as a model for responsive, ecosystem-focused governance in high-activity marine zones worldwide.

4.2. Coastal Benthic Substrate Classification Using Machine Learning

Machine learning holds significant potential in marine habitat mapping, as shown by Labbé-Morissette et al. [13] in their comparative study of the application of supervised and unsupervised models to classify benthic substrates. By employing multibeam echosounder data from the St. Lawrence Estuary, this study not only advances ecological monitoring methods but also provides essential tools for resource management and conservation in coastal habitats. The findings reveal that supervised and unsupervised learning models both offer advantages, with Gaussian mixture models excelling in terms of their efficiency and classification accuracy.

5. Conclusions

This Special Issue of *Geomatics* presents an array of research on ocean mapping and nautical cartography with compelling results, reflecting the significant progress made in this field within recent years. The studies herein demonstrate that innovations in data processing, automation, high-resolution mapping, and ecological planning are not only advancing scientific understanding but are also improving the practical applications of such strategies across the maritime and environmental sectors. By addressing longstanding challenges such as data accuracy, coverage limitations, and the integration of historical datasets, these advancements are paving the way for a more comprehensive, accessible, and sustainable approach to marine spatial planning [14–16].

Future research into ocean mapping will likely continue to combine traditional data sources with emerging technologies like machine learning and satellite geophysics [17,18]. This convergence of technologies is essential in handling the immense and varied data required to map, monitor, and protect ocean ecosystems effectively. As the demand for accurate, high-resolution seafloor data grows, particularly in the context of climate change and resource management, the approaches presented in this Special Issue will serve as models for future research, policy-making, and applied marine sciences. Ultimately, the strides made in this Special Issue exemplify the need for a forward-thinking approach to oceanography, in which data-driven interdisciplinary efforts enhance our capacity to safeguard and sustainably manage the world's oceans.

Conflicts of Interest: The authors declare no conflicts of interest.

References

1. Winther, J.-G.; Dai, M.; Rist, T.; Hoel, A.H.; Li, Y.; Trice, A.; Morrissey, K.; Juinio-Meñez, M.A.; Fernandes, L.; Unger, S.; et al. Integrated ocean management for a sustainable ocean economy. *Nat. Ecol. Evol.* **2020**, *4*, 1451–1458. [CrossRef] [PubMed]
2. Mayer, L.; Jakobsson, M.; Allen, G.; Dorschel, B.; Falconer, R.; Ferrini, V.; Lamarche, G.; Snaith, H.; Weatherall, P. The Nippon Foundation—GEBCO Seabed 2030 Project: The Quest to See the World's Oceans Completely Mapped by 2030. *Geosciences* **2018**, *8*, 63. [CrossRef]
3. Mayer, L. Uncrewed surface systems facilitating a new era of global ocean exploration. *Int. Hydrogr. Rev.* **2023**, *29*, 42–55. [CrossRef]
4. Ryu, P.; Brown, D.; Arsenault, K.; Cho, B.; March, A.; Ali, W.H.; Charous, A.; Lermusiaux, P.F.J. A Wide-Area Deep Ocean Floor Mapping System: Design and Sea Tests. *Geomatics* **2023**, *3*, 290–311. [CrossRef]
5. Younkin, E.; Umfress, S.H. Performing a Sonar Acceptance Test of the Kongsberg EM712 Using Open-Source Software: A Case Study of Kluster. *Geomatics* **2022**, *2*, 540–553. [CrossRef]

6. Masetti, G.; Faulkes, T.; Wilson, M.; Wallace, J. Effective Automated Procedures for Hydrographic Data Review. *Geomatics* **2022**, *2*, 338–354. [CrossRef]
7. Le Deunf, J.; Schmitt, T.; Keramoal, Y.; Jarno, R.; Fally, M. Automating the Management of 300 Years of Ocean Mapping Effort in Order to Improve the Production of Nautical Cartography and Bathymetric Products: Shom's Téthys Workflow. *Geomatics* **2023**, *3*, 239–249. [CrossRef]
8. Rodriguez-Perez, D.; Sanchez-Carnero, N. Multigrid/Multiresolution Interpolation: Reducing Oversmoothing and Other Sampling Effects. *Geomatics* **2022**, *2*, 236–253. [CrossRef]
9. Masetti, G.; Andersen, O.; Andreasen, N.R.; Christiansen, P.S.; Cole, M.A.; Harris, J.P.; Langdahl, K.; Schwenger, L.M.; Sonne, I.B. Denmark's Depth Model: Compilation of Bathymetric Data within the Danish Waters. *Geomatics* **2022**, *2*, 486–498. [CrossRef]
10. Masetti, G.; Granding, P.S.; Langdahl, K.; Schwenger, L.M.; Kristmar, K.V. Denmark's Depth Model version 2.0—Improved compilation of bathymetric data within the Danish waters. *Int. Hydrogr. Rev.* **2024**, *30*, 50–63. [CrossRef]
11. Lemenkova, P. Seafloor and Ocean Crust Structure of the Kerguelen Plateau from Marine Geophysical and Satellite Altimetry Datasets. *Geomatics* **2023**, *3*, 393–426. [CrossRef]
12. Contarinis, S.; Nakos, B.; Pallikaris, A. Introducing Smart Marine Ecosystem-Based Planning (SMEP)—How SMEP Can Drive Marine Spatial Planning Strategy and Its Implementation in Greece. *Geomatics* **2022**, *2*, 197–220. [CrossRef]
13. Labbé-Morissette, G.; Leclercq, T.; Charron-Morneau, P.; Gonthier, D.; Doiron, D.; Chouaer, M.-A.; Munang, D.N. Classification of Coastal Benthic Substrates Using Supervised and Unsupervised Machine Learning Models on North Shore of the St. Lawrence Maritime Estuary (Canada). *Geomatics* **2024**, *4*, 237–252. [CrossRef]
14. Lecours, V.; Oxton, A.; Khor, D.; Tiplea, J. High-Resolution Ensemble Modelling of Coral Distributions in the Northern Gulf of Mexico Based on Geomorphometry: Coral Diversity and Benthic Habitat Fragmentation From Oil and Gas Infrastructure to Inform Spatial Planning. *Aquat. Conserv. Mar. Freshw. Ecosyst.* **2024**, *34*, e4234. [CrossRef]
15. Sowers, D.C.; Mayer, L.A.; Masetti, G.; Cordes, E.; Gasbarro, R.; Lobecker, E.; Cantwell, K.; Candio, S.; Hoy, S.; Malik, M.; et al. Mapping and Geomorphic Characterization of the Vast Cold-Water Coral Mounds of the Blake Plateau. *Geomatics* **2024**, *4*, 17–47. [CrossRef]
16. Church, I.W. Multibeam Sonar Ray-Tracing Uncertainty Evaluation from a Hydrodynamic Model in a Highly Stratified Estuary. *Mar. Geod.* **2020**, *43*, 359–375. [CrossRef]
17. Elmore, P.; Calder, B.; Petry, F.; Masetti, G.; Yager, R. Aggregation Methods Using Bathymetry Sources of Differing Subjective Reliabilities for Navigation Mapping. *Mar. Geod.* **2023**, *46*, 99–128. [CrossRef]
18. Parrish, C.E.; Magruder, L.; Herzfeld, U.; Thomas, N.; Markel, J.; Jasinski, M.; Imahori, G.; Herrmann, J.; Trantow, T.; Borsa, A.; et al. ICESat-2 Bathymetry: Advances in Methods and Science. In Proceedings of the OCEANS 2022, Hampton Roads, VA, USA, 17–20 October 2022; pp. 1–6.

geomatics

Project Report

A Wide-Area Deep Ocean Floor Mapping System: Design and Sea Tests

Paul Ryu [1,*], David Brown [1], Kevin Arsenault [1], Byunggu Cho [1], Andrew March [1,*], Wael H. Ali [2], Aaron Charous [2] and Pierre F. J. Lermusiaux [2,*]

1 Advanced Undersea Systems and Technology Group, MIT Lincoln Laboratory, Lexington, MA 02421, USA
2 Department of Mechanical Engineering, Center for Computational Science and Engineering, Massachusetts Institute of Technology, Cambridge, MA 02139, USA
* Correspondence: ryu@ll.mit.edu (P.R.); amarch@ll.mit.edu (A.M.); pierrel@mit.edu (P.F.J.L.)

Abstract: Mapping the seafloor in the deep ocean is currently performed using sonar systems on surface vessels (low-resolution maps) or undersea vessels (high-resolution maps). Surface-based mapping can cover a much wider search area and is not burdened by the complex logistics required for deploying undersea vessels. However, practical size constraints for a towbody or hull-mounted sonar array result in limits in beamforming and imaging resolution. For cost-effective high-resolution mapping of the deep ocean floor from the surface, a mobile wide-aperture sparse array with subarrays distributed across multiple autonomous surface vessels (ASVs) has been designed. Such a system could enable a surface-based sensor to cover a wide area while achieving high-resolution bathymetry, with resolution cells on the order of 1 m^2 at a 6 km depth. For coherent 3D imaging, such a system must dynamically track the precise relative position of each boat's sonar subarray through ocean-induced motions, estimate water column and bottom reflection properties, and mitigate interference from the array sidelobes. Sea testing of this core sparse acoustic array technology has been conducted, and planning is underway for relative navigation testing with ASVs capable of hosting an acoustic subarray.

Keywords: sparse aperture MIMO sonar; deep ocean bathymetry; precise relative navigation; distributed sensing; ocean modeling; path planning

Citation: Ryu, P.; Brown, D.; Arsenault, K.; Cho, B.; March, A.; Ali, W.H.; Charous, A.; Lermusiaux, P.F.J. A Wide-Area Deep Ocean Floor Mapping System: Design and Sea Tests. *Geomatics* **2023**, *3*, 290–311. https://doi.org/10.3390/geomatics3010016

Academic Editors: Giuseppe Masetti, Ian Church, Anand Hiroji and Ove Andersen

Received: 29 January 2023
Revised: 10 March 2023
Accepted: 17 March 2023
Published: 22 March 2023

1. Introduction

As the ability to map large sections of the seafloor has increased over time, so has this technology's impact on a broad range of scientific disciplines. For example, recent advances in coverage rate and resolution of undersea sensing capabilities have inspired new methods for identifying and tracking hazardous geological processes and vulnerable ecosystems and habitats [1–3]. Large-scale high-resolution mapping can enable safe navigation for submersibles, improved climate modeling, infrastructure monitoring, and disaster response efforts such as locating missing objects. However, the difficulty of recovery tasks in deep and wide areas of the ocean has been highlighted by recent tragedies such as the disappearance of Malaysia Airlines flight 370 in the South Indian Ocean [2,4].

A significant challenge in continuing to advance seafloor mapping is that no technology exists in water deeper than 1 km to obtain meter-scale resolution maps of the seafloor from the ocean surface [4–6]. Current technology capable of finding human-made objects on the seabed or identifying bathymetric features at meter-scale has a maximum range of less than 1 km through the water [7,8]. Because the ocean has an average depth of 3.7 km, high-resolution mapping for the vast majority of the sea requires sending a vehicle underwater, typically within a few hundred meters of the seafloor. Deploying undersea vehicles, even uncrewed ones, to the deep ocean is a time-consuming and expensive endeavor, severely limiting the area such vehicles can map [9–11]. Therefore, wide-area, deep ocean surveys are currently forced to accept lower-resolution goals that surface-vessel-based multibeam sonars can achieve. For instance, the Seabed 2030 project [12] used 400 m × 400 m grids as the baseline seabed cell resolution for water depths from 3 to 5.75 km.

Surface-vessel-based multibeam sonars can map the seabed with a fast coverage rate compared to sonars mounted on undersea vehicles but at a significantly lower resolution [7,13]. To employ sonar ranging through a deep water column, high-amplitude transducers operating at relatively low center frequency must be selected to reduce transmission loss. The selection of frequency determines the maximum range of a sonar array and an upper bound of the resolution set by diffraction limit [14]. Therefore, low-frequency sonar operation that can reach the deep ocean typically yields low-resolution imagery. Each pixel on the map generated by a surface-vessel-based multibeam sonar such as Kongsberg EM302 represents approximately the size of a football field on the deep seabed. By contrast, higher frequency sonars on vehicles that operate closer to the sea bottom, such as Kongsberg EM2040, can achieve a 1 m resolution, resulting in 10,000× more detailed imagery.

This work aims to develop and demonstrate a cost-effective technology for simultaneous high-resolution and rapid seafloor mapping in deep water. To generate comparable resolutions to the high-frequency systems that operate near the bottom of the ocean, a surface-based sonar must have an aperture approximately 100× the size of those currently in use. However, the sonar arrays installed on existing large ocean mapping ships are already at the maximum size supported by their hulls [12,15]. To achieve a sonar array that is both cost-effective and 100× larger than a ship hull, our design concept uses a small swarm of widely spaced ASVs, each hosting a small local sonar array, as shown in Figure 1.

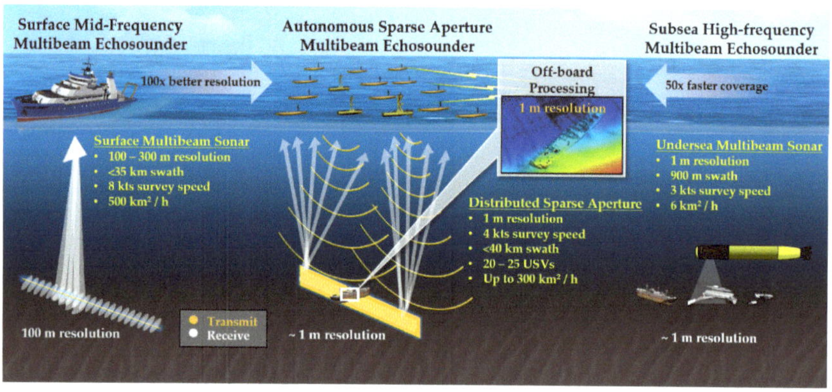

Figure 1. The autonomous sparse-aperture concept offers a potential 100-fold improvement in resolution compared with existing surface ship multibeam sonar, or a potential 50-fold increase in coverage rate compared with subsea sonar.

In this concept, ASVs would work collaboratively to create a large effective array aperture, spanning hundreds of meters. However, for cost reasons, only a limited number of ASVs would be used to synthesize the array, on the order of tens of ASVs separated tens of meters from each other in a random pattern. With this sparse spatial sampling, the array achieves a lower signal-to-noise ratio (SNR) and is less able to reject noise coming from undesired directions because of high sidelobes [14]. However, these challenges can be mitigated using novel signal processing tools and acoustic propagation and imaging algorithms, including Bayesian and artificial intelligence (AI) schemes, enabling sparse arrays to achieve higher resolution. In addition, the ocean sound speed field needs to be estimated through the full water column, and the motions of the ASVs need to be measured very precisely. A surface-based sparse sonar system will need to be integrated with real-time, multiresolution probabilistic ocean field estimation, optimal ASV path planning and coordination, and Bayesian inversions and machine learning of ocean physics and acoustics fields [16–19].

Deploying several copies of this autonomous swarm of ASVs to map the deep ocean seabed at high resolution would enable data collection that could revolutionize our understanding and modeling of the ocean. The area coverage rate—estimated from sensor swath and survey speed—currently achievable for high-resolution mapping with an autonomous underwater vessel is approximately 6 km^2/h, whereas a surface vessel maps at more than 50× that rate, albeit only with low resolution. Efforts are underway to incrementally improve the efficiency of undersea-vehicle-based surveys—for instance, the recently developed PISCES system employs both surface and undersea sensors in a novel bistatic configuration [20]. Our system processes bistatic target reflections with a large sparse aperture operating from the surface and thus is capable of achieving a much greater coverage rate.

Autonomous systems enable cost-effective, extended-duration data collections that can map large areas of the deep ocean bottom. For a sparse-aperture system, the ocean field estimation, ASV path planning and coordination, Bayesian and AI inversions, and signal processing could be completed either onboard the ASVs or remotely. For example, the ASVs could be deployed from a ship, controlled and guided remotely, and left to map the seabed (including searching for a particular object on the seabed) for months until maintenance is needed. All collected data would be returned through a satellite link, and a high-resolution map of the seabed could be generated remotely.

The long-term goal is to demonstrate that a small swarm of ASVs can achieve high-resolution maps of the seabed from the ocean surface. This work addresses two specific objectives:

1. To demonstrate that a sonar array composed of multiple collaborative ASVs can produce a high-resolution map of the seabed in deep water with precise relative positioning knowledge;
2. To demonstrate that a single ASV working in conjunction with a large sonar array can achieve all position, timing, and acoustic signal requirements to work effectively as part of the large array.

Our teams at the Lincoln Laboratory (LL) and Massachusetts Institute of Technology (MIT) have started working toward achieving these two goals. In particular, design studies and risk-reduction efforts for the overall project have been performed, including an analysis of the concept; a demonstration of a sparse aperture in a tank; the development of a low-cost, ASV-sized sonar array; sea testing of a large sonar array; the fabrication of a first-generation ASV; and initial relative navigation algorithm development. Figure 2 shows an overview of the technological development phases. This paper describes the design and specifications of our ocean floor mapping system and the techniques used for data acquisition, signal processing, and imaging. We illustrate the system's capabilities using a proof-of-principle tank test at MIT LL Autonomous Systems Development Facility and an ocean test in Boston Harbor. Furthermore, we highlight our current efforts toward building an autonomous sparse-aperture sonar onboard multiple ASVs and precision-relative navigation system for the ASVs.

The rest of the article is organized as follows. In Section 2, we outline our system design and present the signal-processing chain and imaging algorithms used to obtain 3D seafloor maps. We provide validation results from a tank test using a proof-of-principle sonar in Section 3. In Sections 4 and 5, we describe an application of our system in producing a 3D reconstruction of a sunken barge near Boston Harbor and current efforts to develop precision-relative navigation for ASVs, respectively. In Section 6, we discuss overall system design and performance, and our future project plans are provided in Section 7. Lastly, we discuss environmental impacts in Appendix A and outline our algorithm to reduce sidelobe effects in sparse array data in Appendix B.

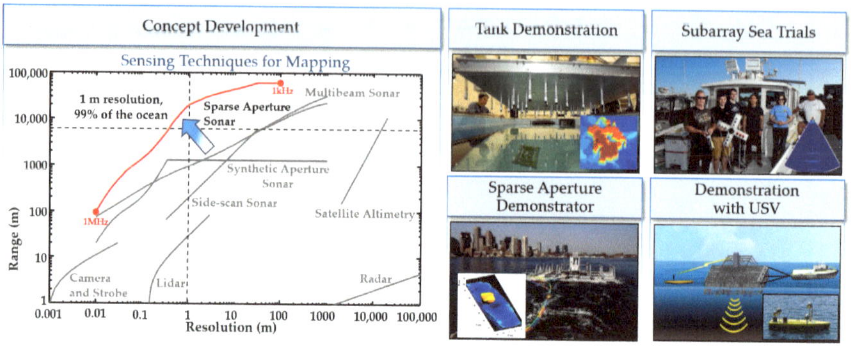

Figure 2. Progression from concept development to sea testing of the sparse aperture.

2. System Overview

The motivation for developing a large acoustic aperture comes from trade-offs between imaging distance, resolution, operating frequency, and aperture size. To illustrate these trade-offs, consider three scenarios with different aperture sizes and resolution requirements. In Figure 3, the blue curves represent the link budget—the maximum imaging distance as a function of frequency given a maximum sound pressure level (186 dB re 1 μPa for marine life safety) and the minimum required SNR. This link budget limit does not change significantly in the three scenarios. The red and green curves are the range resolution and the diffraction limits for each system, respectively. To achieve 100× better range and cross-range resolution, the signal bandwidth needs to be wider and the wavelength needs to be shorter, so both limits shift to the right by two orders of magnitude.

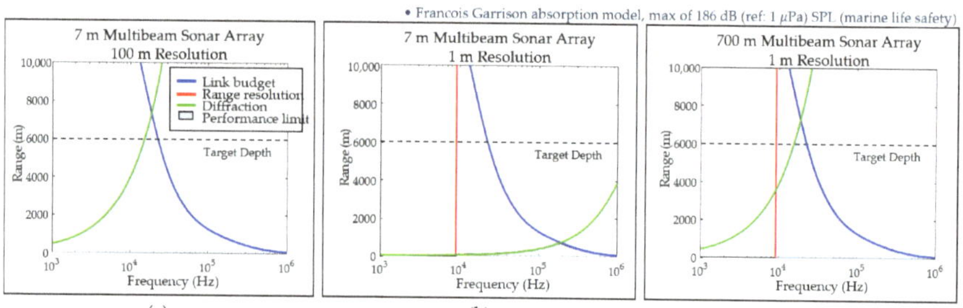

(a) (b) (c)

Figure 3. The performance-bound trade-off between imaging distance, resolution, frequency, and aperture. (**a**) A 7 m sonar array can achieve 100 m resolution at the target depth of 6 km. (**b**) A resolution of 1 m is possible, but the range is limited to 1 km. (**c**) A large 700 m aperture sonar can achieve 1 m resolution at the target depth and beyond.

The shaded areas indicate possible operating regions for the sonar systems. For instance, when operating near 20 kHz, a 7 m sonar array can achieve a 100 m resolution to at least the target depth of 6 km, a depth greater than 99% of the world's oceans' depths. However, to achieve 1 m resolution, the same sonar array can only reach slightly over 1 km. Therefore, the solution to higher resolution at longer distances is to increase the diffraction limit of the system without increasing frequency, thereby requiring the sonar aperture to be significantly increased. For example, a system using a 100× larger array can achieve a 1 m resolution while operating beyond the target depth of 6 km.

A larger sonar aperture increases the diffraction limit of the sonar and enables high-resolution imaging from longer ranges. Traditionally, sonar apertures are populated with transducers at half-wavelength spacing. For very large apertures, this approach would be prohibitive in terms of size, weight, power requirements, and cost. Therefore, we designed a large aperture sonar system with a sparsely populated 2D array of transmit and receive elements.

To demonstrate this concept, we implemented a sparse-aperture multiple-input, multiple-output (MIMO) sonar system that comprises pulse generators, a sparse transmit and receive array, multichannel signal conditioners, and signal converters, as shown in Figure 4. The pulse generators have calibration loopback outputs, which are used as a reference to compensate for time jitter and time drift. Receive elements have flat sensitivity over the operating frequency band. Signal conditioner consists of single-ended-to-differential converter for reducing the impact of common-mode noise pickup and harmonics, low-noise amplifiers for improving dynamic range and sensitivity, and anti-aliasing low-pass filter.

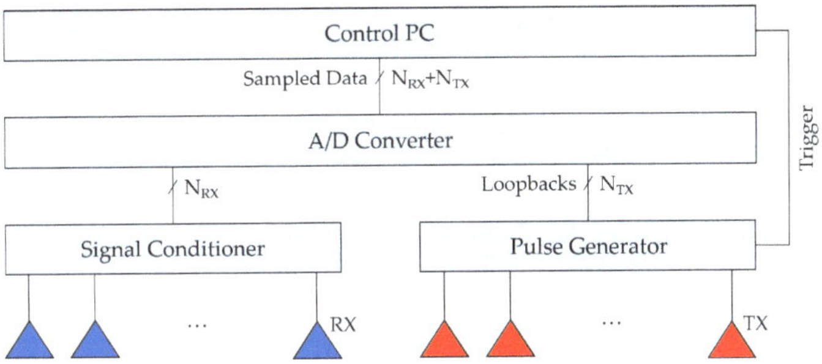

Figure 4. A block diagram of the sparse sonar data acquisition system for N_{RX} receivers (blue) and N_{TX} transmitters (red).

According to Huygens' principle, combining radiated pulses from multiple transmitters to perform space-time focusing to improve array performance is possible [21]. However, this method requires transmitters to generate identical waveforms while maintaining precise positioning and time synchronization. Implementing such an approach for the distributed ASV array would significantly increase the cost and hardware complexity, so the concept of transmit focusing was not included in the current system design.

To achieve orthogonality with MIMO, we adopted a time-division multiple access (TDMA) transmission scheme in which transmitted waveforms share the same frequency band but at different time slots. During each time slot, a single transmitter insonifies the area under the array with a short pulse, and all sparse receive elements are used to sample the scattered wavefield. This sparse MIMO array significantly reduces the number of required transducers and cost but at the expense of less array gain and increased sidelobe level. Therefore, specialized signal processing and image reconstruction algorithms are needed to extract maximal information from the data and to optimize the sparse array performance.

2.1. Signal-Processing Chain

A custom signal-processing chain was developed to condition the data for the image reconstruction step. The first step in the signal-processing chain is to check the health of all source transducers and receive hydrophones. Any sources that do not transmit or any dead or noisy transducers are excised, so they will not be included in the processing. The next step is to match the filter receiving data by convolving it with a replica of the transmitted pulse from each source. Matched filtering increases the temporal resolution of the data and improves the SNR by suppressing noise through coherent integration. Because the waveforms used in the ocean testing sonar system are generated on-demand in hardware and to provide information on the exact time of transmission, the transmitted signals are also recorded on separate loopback channels. The source transducers typically cannot transmit the full frequency extent of the generated waveforms (i.e., they have significant roll-off away from their peak resonant frequency), so a bandpass filter is also applied to suppress the artifact.

The final step of the signal-processing chain is calibrating the data, as shown in Figure 5, by time-aligning the data across receive channels and refining the estimates of array element locations and sound speed needed for image processing. Many hydrophones made up the sonar array used in the ocean testing, which required 12 separate 24-channel audio interfaces (i.e., analog-to-digital converters). Because of this requirement, slight timing differences can exist between channels connected to different audio interfaces. A broadband, pseudo-random noise signal was generated and injected into one channel on each audio interface. These channels were then used to estimate (using time correlation) and correct the relative time offsets between channels on different audio interfaces. The synchronized data were then used to refine the estimates of the transducer and hydrophone locations and the sound speed in the water. The refinement was completed using iterative optimization routines to maximize the total energy at each receiver after summing over the one-way (source-to-receiver or "direct path") signal from all sources. These refined estimates are subsequently used to perform range migration in the image processing step.

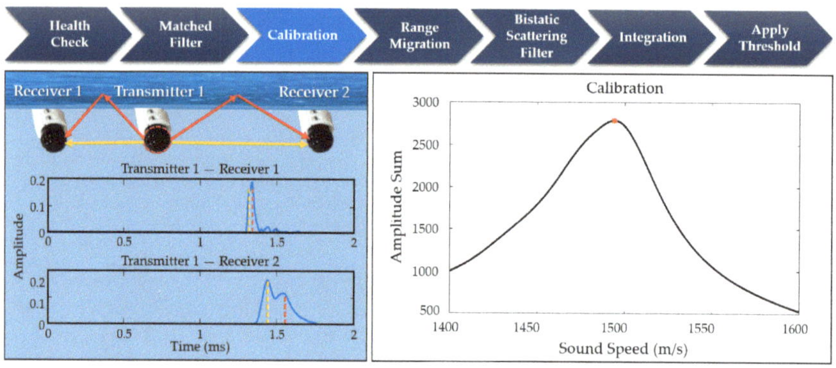

Figure 5. The calibration approach finds optimal sensor positions and sound speed by modeling direct path (orange) and surface bounce path (red) contributions in the matched filtered data (blue).

2.2. Imaging Algorithms

Following geometry processing, data filtering, and calibration, the next step is to define a reconstruction grid underneath the array to cover the search area. We adopted range migration approach to construct the 3D image of the ocean floor because of the advantages this approach offers in computing a high-resolution image at a relatively low computation cost, making it well-suited for near real-time operations. Specifically, we implemented two popular imaging algorithms: diffraction stack [22] and Kirchhoff migration, which was first intuited in 1954 by Hagedoorn [23] and has been refined significantly since then for 2D and 3D applications in acoustic, seismic, and electromagnetic wave imaging [24–26]. Both imaging algorithms can be understood using the following example. If a point in our binning grid was located on the boundary of the object we are interested in imaging, as shown in Figure 6, then the outgoing wave from the transmitter reflects at that point and propagates back to one of the receivers. The pressure signal measured at the receiver will then have a maximum at an instant corresponding to the travel time from the transmitter to the voxel and back to the receiver. Integrating pressure measurements from all receivers at corresponding travel times will show a strong reflectivity value if a voxel lies on the object's surface.

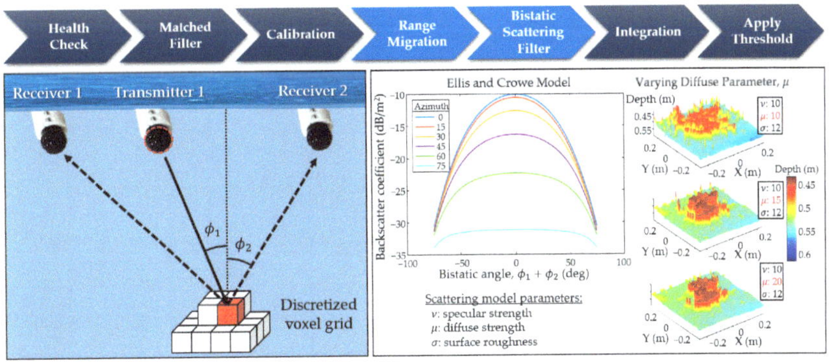

Figure 6. Illustrations of ray tracing in range migration, and the effect of varying the diffuse parameter (μ) in Ellis and Crowe bistatic scattering filter application.

Because of our sparse aperture and the high-resolution requirements, the migrated image suffers from significant sidelobes. To address this limitation, we implemented scattering models to filter the reflected signals at each voxel. Our implementation included models for Lambert's scattering [21], Phong diffuse and specular reflections [27], and the more general Ellis and Crowe bistatic scattering model [28], shown in Figure 6. The imaging and scattering algorithms were implemented for computation using MATLAB, which is a software package that can use graphics processing units to accelerate the computation. For instance, computing a high-resolution 3D image of a 0.5 m $\times$ 0.5 m $\times$ 0.2 m region on a 501 $\times$ 501 $\times$ 201 grid (i.e., $\sim$50 million voxels at $\sim$1 mm resolution) takes around 1.27 min on a NVIDIA RTX A6000 48 GB GDDR6 unit.

The last step in our approach corresponds to visualizing the migrated image. We use a box filter to integrate the energy in a small box of voxels into one larger, lower-resolution voxel, which helps reduce the SNR. In addition, we compare the amplitudes of the voxels and set a threshold for "filled" or "empty"; typically, we have been using 14–18 dB above the minimum energy to be considered material instead of water.

Altogether, the imaging algorithm can be summarized as follows:

1. Range migration—fill the volume underneath the array with voxels and estimate bistatic travel times for each voxel, then sum the matched filter amplitudes at that voxel from all the transmit receiver combinations.
2. Scattering filter—use the scattering angle to filter the return strengths in each voxel.
3. Integration—sum a cube of voxels into one larger voxel, essentially taking 3 $\times$ 3 $\times$ 3 or 15 $\times$ 15 $\times$ 15 voxels and combing all the energy in those voxels into one larger, lower-resolution voxel.
4. Thresholding—compare the amplitudes of the voxels and threshold for "filled" vs. "empty" voxels.

3. Sparse-Aperture Sonar: Scaled Tank Test

To demonstrate that a large sparse-aperture sonar could be made and the sidelobe issue could be mitigated by appropriate array design and signal processing, we built a proof-of-principle sonar in a test tank. The sonar consists of 43 transducers that operate at 200 kHz across a 1.5 m aperture. The array is effectively 200 wavelengths in each direction, and with only 43 transducers, it is very sparse. Figure 7 shows a picture of the sparse sonar array above the test tank and example imagery.

In preparation for the oceangoing system, given that acoustic projectors are significantly more expensive than hydrophones ($\sim$20$\times$ more), the system was biased toward many "receive" channels and few "transmit" channels. All beamforming and timing are performed on the receive side. The system has a custom data acquisition system that samples at 1.67 Msps, and all transmitters have loopbacks.

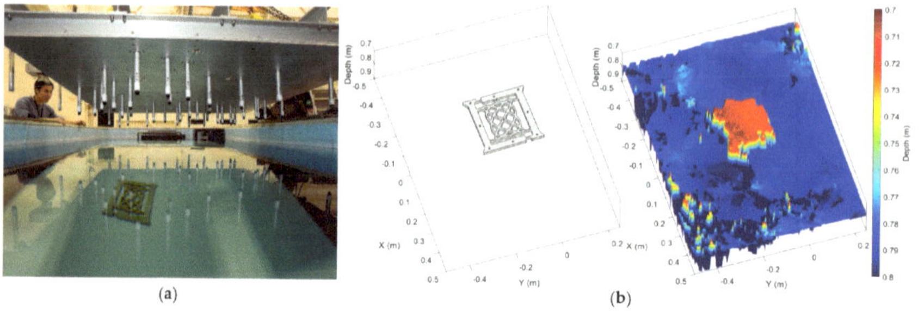

Figure 7. (**a**) The 200 kHz sparse aperture sonar array above the test tank and (**b**) CAD model of the target and acoustic reconstruction example.

To generate the imagery from the acoustic data collected in the tank, we employed our custom signal-processing chain and imaging algorithms presented in Section 2. Figure 7 shows sample imagery of a 10 cm MIT LL logo positioned 10 cm above the tank bottom. The features of the logo are about 9 mm wide, and the acoustic wavelength is about 8 mm, so resolving the finer details of the logo is not expected. However, the overall extent is readily apparent.

The imaging algorithms were compared to attain the sharpest image. Because of the vast volume of 3D, high-resolution data to be processed and limited computing power, we sought an algorithm that produced suitable images at the minimum possible computational expense. In each of the following cases, after processing the data, we set data thresholds such that voxels with energy less than $\varepsilon_{\text{thresh}}$ dB of the maximum are treated as noise and not shown. We started by comparing the two different algorithms referenced in Section 2.2 in Figure 8: Kirchhoff migration and diffraction stack imaging. We used 0.5 cm resolution for this experiment and smoothed the data with a $3 \times 3 \times 3$ box filter. For Kirchhoff migration and diffraction stack imaging, we set $\varepsilon_{\text{thresh}} = -14$ and -9 dB, respectively. Because Kirchhoff migration incorporates angular information, it produced far superior results to the diffraction stack.

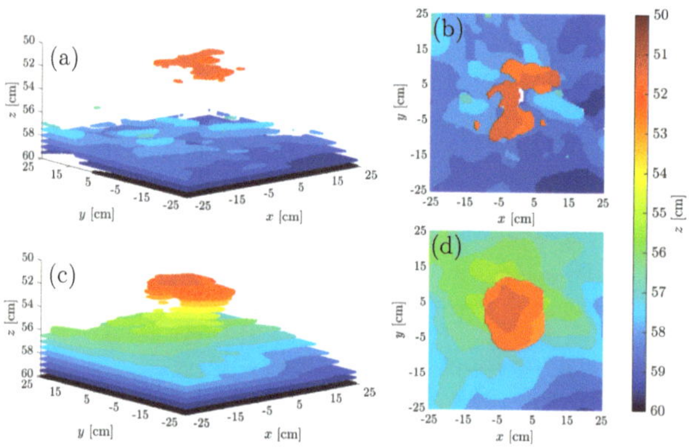

Figure 8. A comparison of the results of Kirchhoff migration (**a**,**b**) and diffraction stack (**c**,**d**) imaging, with (**a**,**c**) showing 3D visualizations and (**b**,**d**) showing a top-down view. Kirchhoff migration successfully localizes the LL logo, while the diffraction stack produces a smoothed image.

Next, we considered the different scattering models discussed in Section 2.2. We processed the data using Ellis' and Crowe's [28], Lambert's [21], and Phong's [27] scattering models, in addition to not using any scattering model at all with thresholds $\varepsilon_{\text{thresh}} = -14, -14, -14$, and -22 dB, respectively. In this experiment, we used a 0.1 cm resolution with a $15 \times 15 \times 15$ box filter to give an equivalent smoothed volume. Figure 9 shows the results of each scattering model. The Ellis and Crowe bistatic model and Phong model outperform the others; however, we note that the scattering models have tunable parameters, and Lambert model results for the tank data can improve with further optimization. Nevertheless, this experiment demonstrates the importance of an appropriate scattering model in 3D imaging.

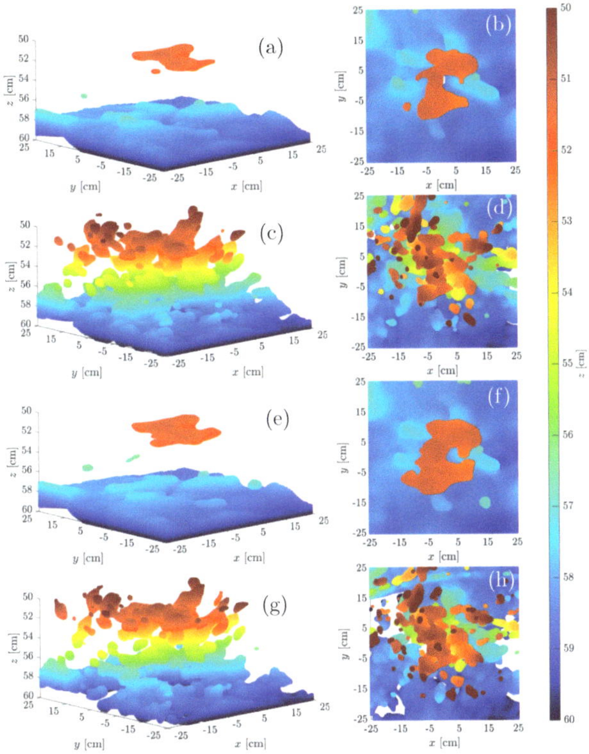

Figure 9. A comparison of the results of Ellis and Crowe [28] (**a**,**b**), Lambert [21] (**c**,**d**), Phong [27] (**e**,**f**), and no scattering model (**g**,**h**). Tuned Ellis and Crowe model and Phong model give the best-resolved images.

Finally, using the Ellis and Crowe bistatic scattering model, we determined at what resolution to process the data. Our goal was to find the minimum resolution such that no discernible details were lost. In Figure 10, we compared 1 cm, 0.5 cm, 0.1 cm, and 0.05 cm resolutions with $\varepsilon_{\text{thresh}} = -14$ dB. In contrast to the previous examples, we did not use a box filter so that none of the fine details in the image would be removed. We find that the 0.1 cm and 0.05 cm images produce similar results. However, voids in imagery start to appear with a coarser resolution setting. At a frequency of 200 kHz and assuming the sound speed of the water is 1500 meters per second, the wavelength of the sound emitted is 0.75 cm. The Nyquist criterion then dictates that we sample at least at a resolution of 0.375 cm. This analysis aligns with our experiment, as 0.1 cm is the maximum voxel size tested that satisfies this criterion and the resolution at which we see the most image details.

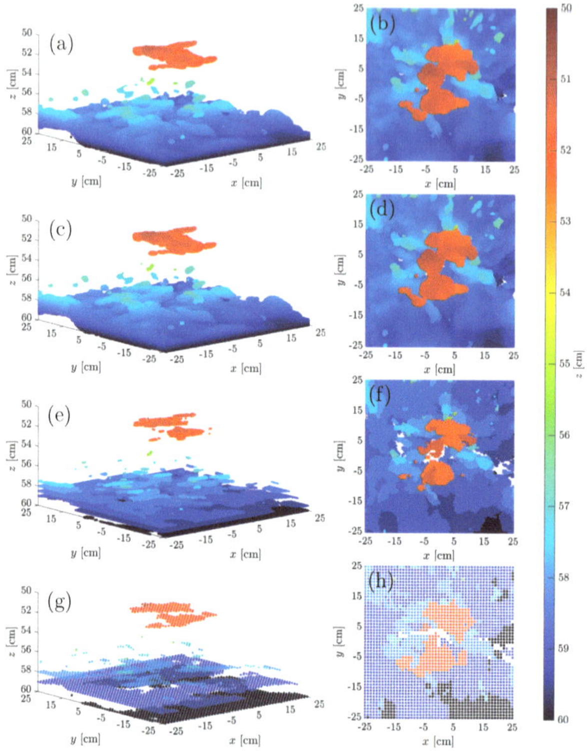

Figure 10. A comparison of the results of 0.05 cm (**a**,**b**), 0.1 cm (**c**,**d**), 0.5 cm (**e**,**f**), and 1 cm (**g**,**h**) resolution imaging. The images lose details at resolutions larger than 0.1 cm, but the finer 0.05 cm resolution image provides no further benefit.

4. Sparse-Aperture Sonar: Ocean Test

The tank testing shows that the sparse-aperture concept itself is viable. However, tank experiments cannot fully simulate the ocean's complex and dynamic acoustic propagation conditions and the relative position variations in distributed arrays resulting from ocean waves and currents. To address the former, we built an oceangoing large sparse-aperture sonar testbed, shown in Figure 11. The metal frame ensures that the hydrophones and transmitters are locked in position and that the array geometry varies by less than a tenth of the wavelength (4 mm), even in rough seas.

(**a**) (**b**)

Figure 11. (**a**) The initial deployment of the oceangoing sparse-aperture sonar. (**b**) The sonar is towed out to the ocean through the Boston Harbor on 17 September 2020.

The frame holds together a sparse array of 6 transmitters and 19 cross-shaped receiving subarrays (with a total of 247 hydrophones). The frame also includes two inertial measurement units and a waterproof server rack that houses the data acquisition system with signal conditioners, signal converters, pulse generators for transmitters, and control PCs. All the sensors and electronics are remotely controllable via Ethernet. When assembled, the array is 8 m × 8 m (24′ × 24′), but it can be separated into three units to fit onto a single 53′ flatbed trailer. The system operates at 33 kHz and is designed for an initial operating depth of 150 m. Figure 12 shows the sparse array layout, range profiles of all transmitter–receiver pairs, and a bathymetry point cloud from pier-side sensor data acquisition testing and validation.

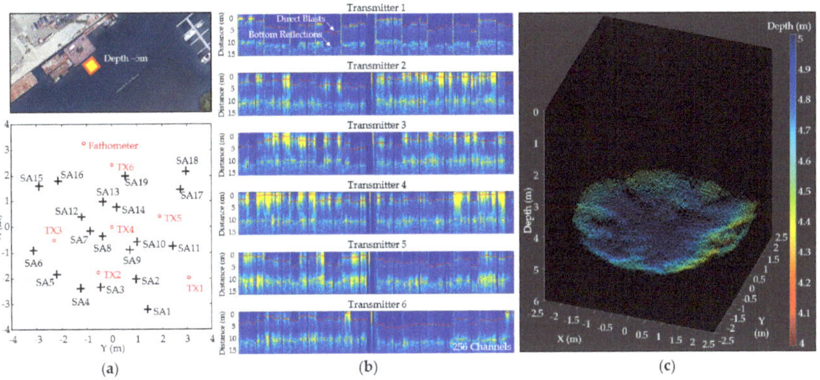

Figure 12. Initial pier-side test results of the oceangoing sparse-aperture sonar. The array layout and range profiles from all transmitter–receiver pairs are shown in (**a**,**b**), respectively. The reconstruction point cloud of the pier bottom is shown in (**c**).

We identified sunken objects near Boston Harbor to image with our sonar array for an initial at-sea demonstration of the sparse-aperture concept. We installed a reference fathometer and ran Humminbird HELIX 7 side-scan sonar to generate reference imagery of the seabed. This side-scan sonar operates at 462 kHz and has a maximum range of 60 m. With 14× the operating frequency of our sparse array, the side-scan sonar is better equipped to resolve small features. Figure 13 shows the reference side-scan imagery and the imagery generated from our sparse sonar array of a sunken barge outside Boston Harbor. The sparse sonar array operates as a multibeam sonar. It can produce a full 3D reconstruction of the seabed, whereas the side scan generates an intensity versus range plot with narrow strips. Because of this difference in image formation techniques, a direct comparison of the two outputs is difficult. Qualitatively, however, the image resolution we are obtaining from our array is similar to that of the side-scan system. A more capable reference multibeam system will be used for further validation and quantitative comparison of the resolution.

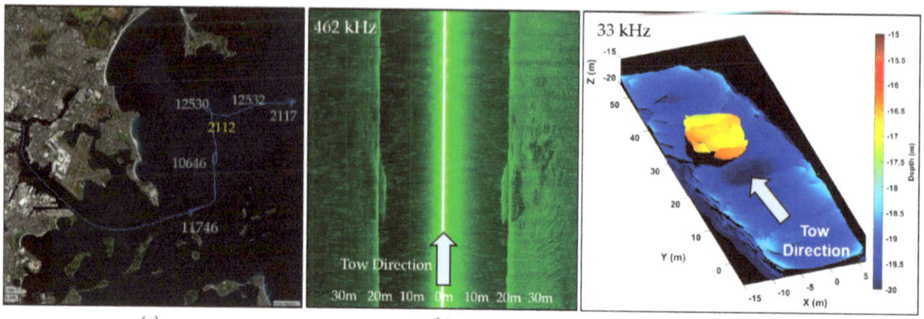

(a) (b) (c)

Figure 13. A comparison of seabed imagery of automated wreck and obstruction information system (AWOIS) site 2112, a sunken barge outside Boston Harbor. The site's geolocation is indicated in (**a**). (**b**,**c**) Show reference imagery from a commercial side-scan sonar and imagery generated with our large sparse array using our initial low-resolution algorithm, respectively.

5. Toward Autonomous Sparse-Aperture Sonar

As we have shown that sparse-aperture imaging of the seabed is possible, we are working to disaggregate the sonar array onto multiple ASVs. The first step is to derive the required position knowledge for all the hydrophones in the array to generate sharp and focused imagery. This position knowledge will serve as a requirement for how well each of the ASVs needs to be tracked. Using our tank demonstration sonar array, we confirmed that if we know the positions of all elements to one-tenth of the acoustic wavelength, we can generate clear imagery. Using our ocean data at 33 kHz and additional modeling, we can estimate the array's performance degradation with element position error.

For example, the wavelength at 33 kHz is 4.5 cm, and from the tank experiments, we would expect sharply focused imagery with a 4.5 mm or less element position error. The power loss with simulated position errors at 33 kHz is shown in Figure 14. We see minimal degradation through the one-tenth wavelength uncertainty and a less than 0.5 dB loss. The 3 dB loss point is around 8 mm, or about one-sixth of a wavelength, and at about one-third of a wavelength, the performance drops precipitously. Therefore, the required position accuracy for hydrophones on the first-generation ASV will be one-tenth of a wavelength, 5 mm, the hope is to develop autofocus techniques to relax this requirement to one-third or one-fifth of a wavelength and reduce the cost of the ASVs.

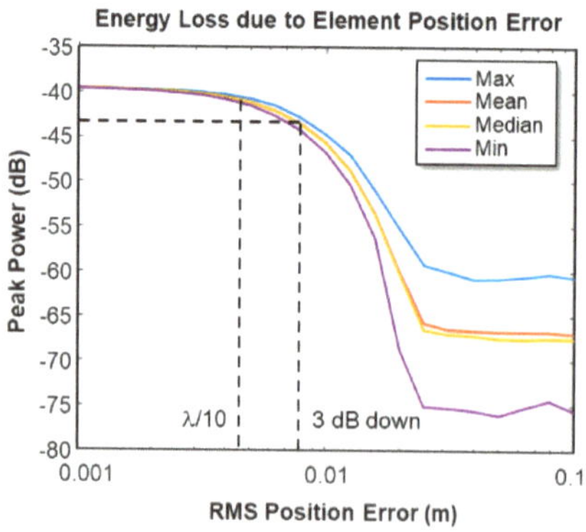

Figure 14. Power loss with position error simulated based on 33 kHz sonar data.

We are working to demonstrate a low-cost, precision-relative navigation capability to achieve 5 mm, 3D relative position accuracy. We performed a trade study evaluating many potential relative navigation options: (1) GPS, (2) inertial measurement, (3) optical tracking, (4) LiDAR, (5) radio-frequency ranging, and (6) high-frequency (automotive) radar. The key performance parameters are sensor field of view, maximum range, measurement accuracy, and cost. We plan to use a fused GPS and inertial measurement with optical and LiDAR tracking. The stereo-optical measurements can be used to generate the most accurate relative measurements; however, this approach requires many cameras and targets, which may exceed cost constraints. LiDAR has excellent coverage and good potential solution accuracy. Figure 15 shows an example of a "dummy" ASV in a tank with GPS and inertial measurements onboard and offboard tracking with an optical sensor and LiDAR. Our current optical system post-processing can measure optical-marker rigid-body edge distances and interior angles with accuracies of 0.33 cm and 0.25°, respectively.

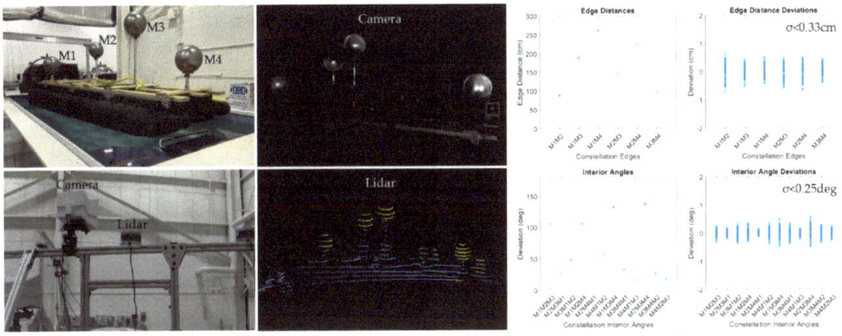

(a) (b) (c)

Figure 15. The in-lab test setup for precision-relative navigation is shown in (**a**). Example data frames from a near-infrared camera and LiDAR are shown in (**b**). Edge distances and interior angles calculated for tracking the ASV are shown in (**c**).

In collaboration with the Woods Hole Oceanographic Institution, we designed a low-cost ASV with a sonar array on the bottom that could have fiducials installed to enable precision tracking. The first of these ASVs has just been deployed and includes a 21-element passive sonar array, fully autonomous control, fused inertial navigation and GPS, and rigid markers for tracking. Pictures from the sea tests are shown in Figure 16. Optical markers are tracked using stereo cameras and LiDARs to generate precise 6-degrees-of-freedom relative poses of the ASVs.

(a) (b)

Figure 16. Initial sea testing of custom ASVs and sensors. Precision optical and LiDAR tracking of the vessel from shore (**a**) and boat (**b**).

6. Discussion: Overall System Design and Performance

In Section 3, we validated our ocean mapping system using a proof-of-principle sonar test in a scaled tank. The tank testing showed that the sparse-aperture concept is viable despite the challenges posed by the tank size causing significant signal interference and multipath effects. We also showcased the application of our imaging algorithms and showed the convergence of reconstructed images as we varied resolution. We compared the reconstructed images from the three scattering models and concluded that the Ellis and Crowe bistatic model provided the best results.

After demonstrating that the sparse-aperture concept is viable, in Section 4 we showcased the application of our system in reconstructing a 3D image of a sunken barge near Boston Harbor. We discussed details of our signal-processing and calibration chain and compared the image we obtained of the barge to that obtained by a commercial side-scan sonar. It should be noted that the imagery from our array was created using an initial algorithm set based on the signal-processing chain we developed for our tank experiment. Many aspects of the signal-processing suite would improve the quality of the imagery. Some of the possible improvements we are currently working on are listed in Table 1. In Appendix A, we discuss environmental impacts and how they can be expected to be limited compared to other systems.

Table 1. Signal processing improvements.

Low Resolution (Existing Algorithm)	Improved Resolution (Future Algorithm)
10 cm voxels	<5 cm voxels
Semi-static motion compensation	Full motion compensation
Range migration with straight propagation	Range migration with ray tracing
Direct blast timing calibration	Autofocus for calibration
Measured sound velocity profile	Sound velocity updated during focusing
Basic bistatic scattering model	Scattering model optimization
Incoherent summing	Coherent processing

A sparse array design introduces higher sidelobes than a fully populated dense array of equal size. A high sidelobe allows acoustic energy from other directions to smear into the look location of the array and degrades the image quality. In Appendix B, we showcase such sidelobe effects and outline our algorithm to reduce them. Specifically, Figures A2 and A3 first confirm that a long inter-array distance introduces high sidelobe artifacts. Figure A4 shows that these sidelobe artifacts intensify for a spotlight-shaped transmit beam. This wide transmit beam illuminates a broader 2D patch of the seafloor than a 1D seafloor strip illumination from a narrow fan-shaped transmit beam and introduces more sidelobe energy leakage. We have developed a coherent sidelobe-reduction algorithm based on alternative projection [14] and the CLEAN algorithm of Hogbom [29]. Our algorithm effectively reduces sidelobe artifacts for a modeled Lambertian seafloor example, as shown in Figure A5. This algorithm is currently under further development, and we are planning performance analyses on both simulated scenarios and real measurements.

On the basis of our current ASV design, tracking system, and acoustic imaging, we have estimated the performance and costs of a fully operational array. Desired resolution of the seabed imagery determines the sonar operating frequency and the number and quality of navigation sensors needed to achieve the one-tenth-wavelength relative positioning accuracy. The accuracy requirement can be relaxed to one-third-wavelength with an autofocus mapping capability.

A conventional deep-water multibeam sonar has about 1° beam and costs around USD 2 million. The sparse array technique will require long-endurance ASVs that cost about USD 75,000 each. At 1 m resolution, using our current GNSS–visual–inertial navigation setup will require installing a precision GNSS receiver capable of real-time kinematics (RTK), inertial measurement unit, and multiple cameras and LiDARs on vehicles. The total cost of the system will be USD 7 million At lesser resolution requirements, such as 10 m, fewer ASVs are needed, and each vehicle can be outfitted with a less exquisite navigation sensor suite. The total cost drops to USD 2 million. An overview of this trade is shown in Figure 17, which shows the Pareto front of cost and resolution for the operational system.

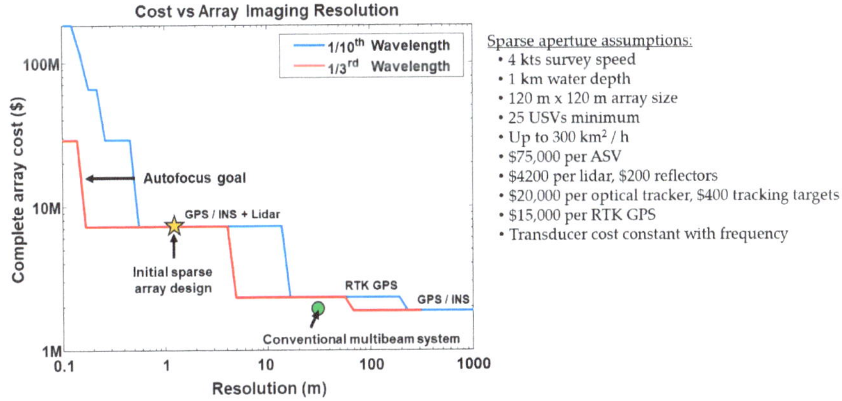

Figure 17. The potential cost of an operational sparse-aperture sonar mapping using at least 25 ASVs and our current image reconstruction, focus, and relative navigation techniques.

7. Future Work

We plan to improve signal-processing and 3D image reconstruction algorithms by incorporating autofocus methods. We could also investigate merging these advances with our Bayesian uncertainty quantification and learning using our dynamically orthogonal stochastic acoustic wave predictions [30,31]. A key aspect of the planned sea tests is the prediction of the ocean fields (e.g., sound speed, currents) to be used in estimating the bathymetry and seabed properties from the acoustic data and some temperature, salinity, and biogeochemical data. We could employ our MIT Multidisciplinary Simulation, Estimation, and Assimilation System (MSEAS) ocean modeling system [32,33], optimal reduced-order theory and schemes for uncertainty quantification, and Bayesian data assimilation and learning [34,35]. We could first predict joint probability density functions (PDFs) of the uncertain ocean and seabed fields, using our new coupled dynamically orthogonal stochastic ocean acoustic partial differential equations [36–39], with extension to multiresolution probabilistic predictions. With these prior PDFs, our Bayesian inference algorithms then assimilate the measured acoustic data and sound speed profiles to jointly estimate posterior PDFs for the ocean fields, bathymetry, and seabed's geoacoustic properties [36,40]. With Bayesian and machine learning [41,42], we can then jointly learn better parameterizations (e.g., for scattering), neural model closures (e.g., for missing or unknown model terms), and even model formulations themselves (e.g., for model complexity).

For the control and coordination of the ASVs, we could utilize our principled optimal path planning and coordination for operations in strong dynamic ocean currents and waves (e.g., [19,43,44]). With this model-based predictive control and path planning coordination, we could guide the ASV headings and speeds to ensure that the ASV remains in an optimal location with respect to the rest of the array. For multiple ASVs, our rigorous path planning and coordination methods could help maintain larger-scale patterns that are optimal for the inference of bathymetry while accounting for ocean currents and waves.

Author Contributions: Conceptualization and methodology, P.R., A.M. and P.F.J.L.; software, P.R., K.A., B.C., A.M., W.H.A. and A.C.; investigation and data curation, P.R., D.B. and A.M.; writing—original draft preparation and writing—review and editing, all authors; supervision and project administration, A.M. and P.F.J.L. All authors have read and agreed to the published version of the manuscript.

Funding: This research is supported by the Under Secretary of Defense for Research and Engineering under Air Force Contract No. FA8702-15-D-0001. Any opinions, findings, conclusions or recommendations expressed in this material are those of the authors and do not necessarily reflect the views of the Under Secretary of Defense for Research and Engineering.

Institutional Review Board Statement: Not applicable.

Informed Consent Statement: Not applicable.

Data Availability Statement: The data presented in this study are available on request from the corresponding authors.

Acknowledgments: We thank Katherine Rimpau of MIT Lincoln Laboratory for assistance with data acquisition and data analysis, Paul Pepin of MIT Lincoln Laboratory for exceptional design of the apparatuses, and our colleagues on the MIT Lincoln Laboratory Autonomous Systems Line and Integrated Systems Line Committees for their helpful technical advice.

Conflicts of Interest: The authors declare no conflict of interest.

Appendix A. Environmental Impacts

The only foreseen potential environmental impact of this system is the acoustic effects on marine life. A typical multibeam echosounder on a surface ship designed for deep water produces a 240 dBuPA@1m sound pressure level (SPL) and has a wide beam of around 140°. Given the expense of ship operations, the sonar is run continuously. Our current rigid sonar array is intentionally power limited such that it does not require permitting to use. The acoustic projectors are standard commercial fathometers modified to play alternative waveforms at reduced maximum output power. They also have a very narrow, 22°, downward-pointing beam. Using 160 dB SPL criteria for Level B harassment, a marine mammal within 10,000 m of the ship would be behaviorally harassed. However, with our current array, a marine mammal would need to be within 8 m and 11° off-vertical from our transducer to be behaviorally harassed. An animal must be underneath the array and within 8 m to be affected by our testing. Overall, the environmental impact of the prototype sparse-aperture array is less than that of other vessel traffic employing a low-frequency fathometer.

For a future operational system using ASVs, the beam width of the transmitters and power level will be increased. The power level will still be less than that of a surface ship multibeam. To mitigate the increase in SPL and beam width, we intend to develop an automated detector of marine mammal sounds and pause testing when marine mammals are suspected to be in the vicinity of the array. The testing pause is much less impactful for an autonomous system than for a large capital ship. Figure A1 compares the rough areas underneath a conventional multibeam, our existing array, and a future potential sparse-aperture operational system in which a marine mammal may be harassed.

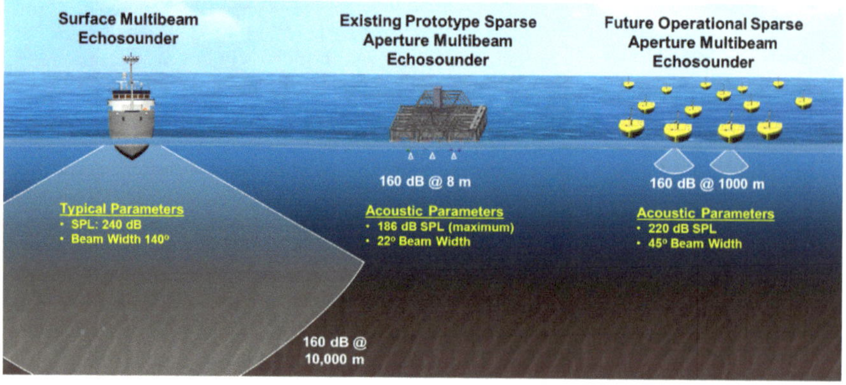

Figure A1. Approximate distance from acoustic sources that exceed 160 dB sound pressure level for a typical ship-based multibeam echosounder, our existing prototype array, and a future first-generation operational system.

Appendix B. High Sidelobe Effects in a Sparse Array Image and Reduction Algorithm Development

This section shows the sidelobe effects in a sparse array image and investigates their cause. We also briefly introduce a potential sidelobe reduction algorithm and demonstrate its effectiveness by simulation.

Appendix B.1. High Sidelobe Effects in a Sparse Array Image

Here, we simulate the sidelobe effects in a sparse array image and compare it with a dense-array image. Figures A2 and A3 show that a sparse array leads to more sidelobe artifacts and degrades the image quality. This outcome is due to sidelobe energy from other seafloor pixels smearing into the array's look position.

A 2D seafloor further degrades image quality, as shown in Figures A3 and A4. A 1D seafloor strip effectively models a fan-shaped transmit beam wide in one horizontal direction and narrow in the other (Figure A3). A 2D seafloor patch models a spotlight-shaped transmit beam wide in both horizontal directions (Figure A4). This spotlight-shaped transmit beam illuminates a relatively wide area of the seafloor and increases energy smearing from the seafloor around the array's look location.

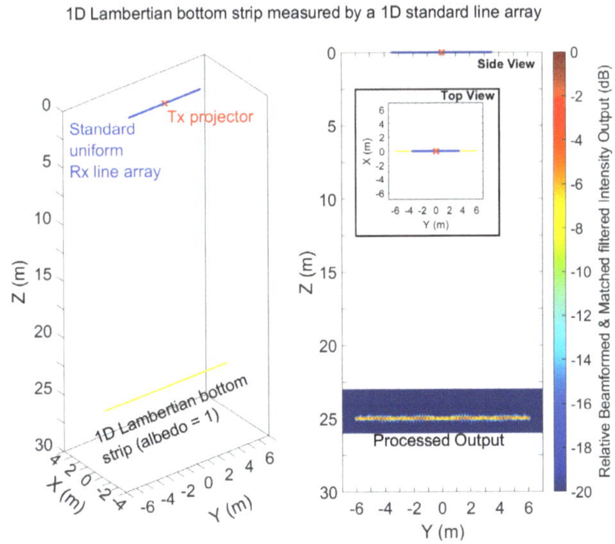

Figure A2. A simulated 1D Lambertian seafloor strip imaged by a dense uniform line receive array. A 0.05 s linear frequency-modulated transmit waveform at 33 ± 3 kHz is projected from a transmit element at the center of the receive array. The received signal at each receive element is matched filtered and then beam-formed for each spatial voxel position. Scattered energy is well-contained within the seafloor region. No random ambient noise is used.

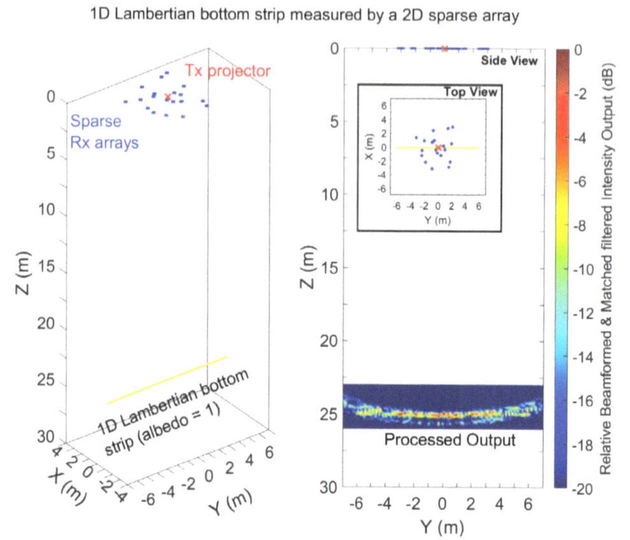

Figure A3. Simulated imaging with a sparse array. The environmental configuration and transmit waveform are the same as those in Figure A2. The geometry of the sparse receive array and the transmitter is shown. The matched filtered and beamforming scattered output energy is spread around the seafloor region because of the sidelobes. Note that a 1D bottom strip models a fan-shaped transmit beam that is narrow in the x direction and wide in the y direction.

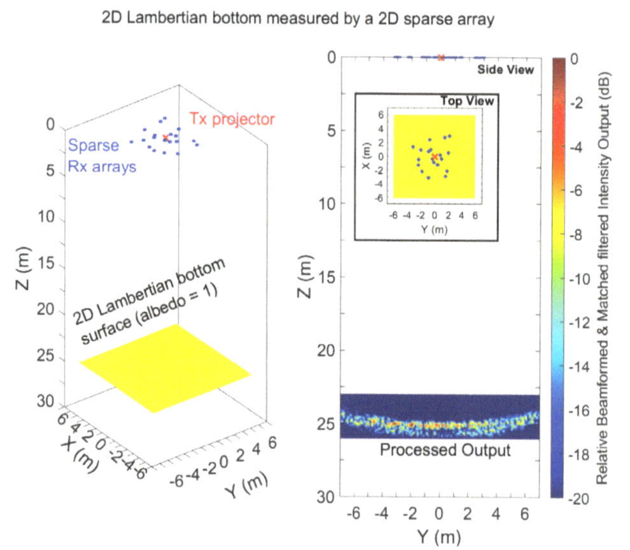

Figure A4. A simulated sparse array image of a 2D Lambertian seafloor. The array configurations are the same as those in Figure A3. A 2D seafloor effectively represents a spotlight-shaped transmit beam that is wide in both x and y directions. The 2D seafloor scattering increases sidelobe artifacts and further degrades image quality.

Appendix B.2. Sidelobe-Reduction Algorithm Development

We are developing algorithms to reduce sidelobe artifacts and improve image quality. We briefly describe the present algorithm and demonstrate its performance for a simple simulated case.

Our algorithm shares a similar philosophy with the alternative projection [14] that is widely used in passive sonar applications. Both algorithms estimate a single-point source and its waveform at each iteration. Alternative projection estimates the signal waveform of a point source as the entire component contained in its estimated direction. Estimating the waveform with a single step can cause energy smearing from nearby sources. Imaging a cluttered region with a sparse array system with typically high sidelobe levels can suffer from energy smearing from other sources or background noise and severely contaminate the waveform estimate. Our algorithm partially estimates the underlying point source's waveform at each iteration. The estimation is less susceptible to contamination by other point sources or background noise, as sidelobe leakage from neighboring sources also decreases as iterations progress.

The iterative scheme we implement is inspired by the CLEAN algorithm of Hogbom [29], which is commonly used for long-baseline astronomical interferometry. However, our deconvolution algorithm is different from the original CLEAN algorithm. The CLEAN algorithm was originally developed based on the van Certtite–Zernike theorem [45,46]. It uses the duality property of the cross-correlation of the electric field measured at two distinct points, i.e., visibility and directional spectral density. The CLEAN algorithm operates on the second-moment statistics and removes sidelobe power artifacts. Our algorithm uses the direct acoustic pressure field and estimates the location and waveform of underlying point sources to remove the sidelobe field artifacts.

Our algorithm starts with matched filtering, followed by beamforming the received signal. This coherent processing is typically performed in the complex baseband for computational efficiency, and, as a result, the raw processed outputs are also complex values. The order of matched filtering and beamforming are interchangeable because both processes are linear. It is, however, computationally efficient to first matched filter the element time series when the number of voxels exceeds the number of "receive" elements. We define the raw complex output of the coherent processing as the "complex dirty map", following the terminology in the original CLEAN algorithm [29]. The algorithm then finds the power peak location in the dirty map and calculates the complex point response function at that peak location. The complex point response function is defined as the "complex dirty beam". The peak height of the complex dirty beam is scaled by 10–25% of the dirty map power peak strength with a phase identical to the complex dirty map peak. This scaled, phase-shifted complex point source at the detected peak position is added to an empty spatial map and forms the "complex clean map". A "complex residual map" is also calculated by subtracting the complex dirty beam of this point source from the initial dirty map. Our algorithm repeats the same procedure, except the updated complex residual map is used instead of the complex dirty map to detect the power peak in consecutive iterations. Iteration continues until the peak power in the complex residual map falls below a desired level—for example, 30 dB below the peak power of the original complex dirty map.

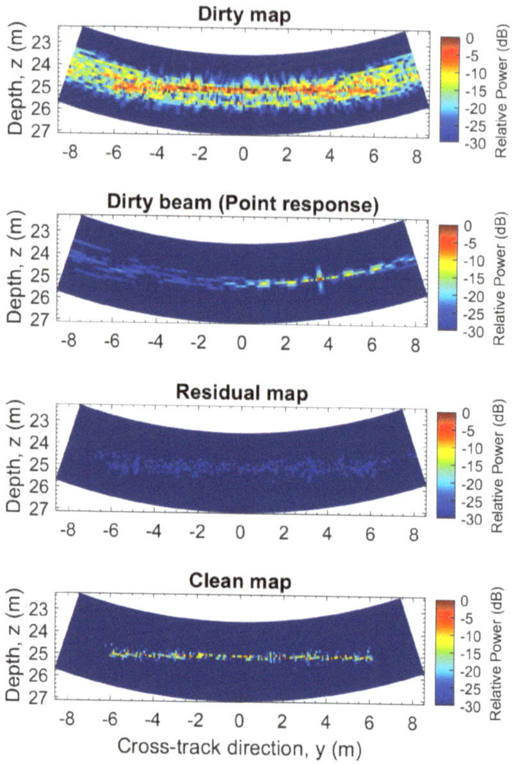

Figure A5. Example of a sidelobe-reduction algorithm for a simulated 1D Lambertian seafloor. The environmental and array configurations are the same as those in Figure A3. The estimated point source locations are well contained within the seafloor region, as shown by the final clean map after 1000 iterations.

References

1. Rengstorf, A.M.; Mohn, C.; Brown, C.; Wisz, M.S.; Grehan, A.J. Predicting the distribution of deep-sea vulnerable marine ecosystems using high-resolution data: Considerations and novel approaches. *Deep Sea Res. Part Oceanogr. Res. Pap.* **2014**, *93*, 72–82. [CrossRef]
2. Picard, K.; Brooke, B.P.; Harris, P.T.; Siwabessy, P.J.; Coffin, M.F.; Tran, M.; Spinoccia, M.; Weales, J.; Macmillan-Lawler, M.; Sullivan, J. Malaysia Airlines flight MH370 search data reveal geomorphology and seafloor processes in the remote southeast Indian Ocean. *Mar. Geol.* **2018**, *395*, 301–319. [CrossRef]
3. Chiocci, F.L.; Cattaneo, A.; Urgeles, R. Seafloor mapping for geohazard assessment: State of the art. *Mar. Geophys. Res.* **2011**, *32*, 1–11. [CrossRef]
4. Picard, K.; Tran, M.; Siwabessy, P.J.; Kennedy, P.; Smith, W. Increased resolution bathymetry in the southeast Indian ocean: MH370 search data. *Hydro Int.* **2017**, *21*, 14–19.
5. Kearns, T.A.; Breman, J. Bathymetry—The art and science of seafloor modeling for modern applications. *Ocean Globe* **2010**, 1–36.
6. Levin, L.A.; Bett, B.J.; Gates, A.R.; Heimbach, P.; Howe, B.M.; Janssen, F.; McCurdy, A.; Ruhl, H.A.; Snelgrove, P.; Stocks, K.I.; et al. Global Observing Needs in the Deep Ocean. *Front. Mar. Sci.* **2019**, *6*, 241. [CrossRef]
7. Wölfl, A.C.; Snaith, H.; Amirebrahimi, S.; Devey, C.W.; Dorschel, B.; Ferrini, V.; Huvenne, V.A.I.; Jakobsson, M.; Jencks, J.; Johnston, G.; et al. Seafloor Mapping—The Challenge of a Truly Global Ocean Bathymetry. *Front. Mar. Sci.* **2019**, *6*, 283. [CrossRef]
8. Hansen, R.E.; Callow, H.J.; Sabo, T.O.; Synnes, S.A.V. Challenges in seafloor imaging and mapping with synthetic aperture sonar. *IEEE Trans. Geosci. Remote Sens.* **2011**, *49*, 3677–3687. [CrossRef]
9. Jalving, B. Depth accuracy in seabed mapping with underwater vehicles. In Proceedings of the Oceans '99—Riding the Crest into the 21st Century, Conference and Exhibition, Conference Proceedings (IEEE Cat. No.99CH37008), Seattle, WA, USA, 13–16 September 1999; Volume 2, pp. 973–978.

10. Wynn, R.B.; Huvenne, V.A.; Le Bas, T.P.; Murton, B.J.; Connelly, D.P.; Bett, B.J.; Ruhl, H.A.; Morris, K.J.; Peakall, J.; Parsons, D.R.; et al. Autonomous Underwater Vehicles (AUVs): Their past, present and future contributions to the advancement of marine geoscience. *Mar. Geol.* **2014**, *352*, 451–468. [CrossRef]
11. Lucieer, V.L.; Forrest, A.L. Emerging mapping techniques for autonomous underwater vehicles (AUVs). In *Seafloor Mapping Along Continental Shelves*; Springer: Berlin/Heidelberg, Germany, 2016; pp. 53–67.
12. Mayer, L.; Jakobsson, M.; Allen, G.; Dorschel, B.; Falconer, R.; Ferrini, V.; Lamarche, G.; Snaith, H.; Weatherall, P. The Nippon Foundation GEBCO seabed 2030 project: The quest to see the world's oceans completely mapped by 2030. *Geosciences* **2018**, *8*, 63. [CrossRef]
13. Jakobsson, M.; Allen, G.; Carbotte, S.; Falconer, R.; Ferrini, V.; Marks, K.; Mayer, L.; Rovere, M.; Schmitt, T.; Weatherall, P.; et al. The Nippon Foundation GEBCO Seabed 2030: Roadmap for Future Ocean Floor Mapping. 2017. The Nippon Foundation GEBCO. Available online https://seabed2030.gebco.net/documents/seabed2030 (accessed on 6 March 2023)
14. Van Trees, H.L. *Optimum Array Processing: Part IV of Detection, Estimation, and Modulation Theory*; John Wiley & Sons: Hoboken, NJ, USA, 2004.
15. Mayer, L.A. Frontiers in seafloor mapping and visualization. *Mar. Geophys. Res.* **2006**, *27*, 7–17. [CrossRef]
16. Robinson, A.R.; Lermusiaux, P.F.J. Prediction Systems with Data Assimilation for Coupled Ocean Science and Ocean Acoustics. In Proceedings of the Sixth International Conference on Theoretical and Computational Acoustics, Honolulu, HI, USA, 11–15 August 2003; Tolstoy, A., Teng, Y., Shang, E.C., Eds.; pp. 325–342.
17. Lermusiaux, P.F.J.; Haley, P.J., Jr.; Yilmaz, N.K. Environmental prediction, path planning and adaptive sampling: Sensing and modeling for efficient ocean monitoring, management and pollution control. *Sea Technol.* **2007**, *48*, 35–38.
18. Wang, D.; Lermusiaux, P.F.J.; Haley, P.J., Jr.; Eickstedt, D.; Leslie, W.G.; Schmidt, H. Acoustically focused adaptive sampling and on-board routing for marine rapid environmental assessment. *J. Mar. Syst.* **2009**, *78*, S393–S407. [CrossRef]
19. Lermusiaux, P.F.J.; Subramani, D.N.; Lin, J.; Kulkarni, C.S.; Gupta, A.; Dutt, A.; Lolla, T.; Haley, P.J., Jr.; Ali, W.H.; Mirabito, C.; et al. A Future for Intelligent Autonomous Ocean Observing Systems. *J. Mar. Res.* **2017**, *75*, 765–813. [CrossRef]
20. Cruz, N.; Abreu, N.; Almeida, J.; Almeida, R.; Alves, J.; Dias, A.; Ferreira, B.; Ferreira, H.; Gonçalves, C.; Martins, A.; et al. Cooperative deep water seafloor mapping with heterogeneous robotic platforms. In Proceedings of the OCEANS 2017—Anchorage, Anchorage, AK, USA, 18–21 September 2017; pp. 1–7.
21. Urick, R.J. *Principles of Underwater Sound*, 2nd ed.; McGraw–Hill: New York, NY, USA, 1975.
22. French, W.S. Two-dimensional and three-dimensional migration of model-experiment reflection profiles. *Geophysics* **1974**, *39*, 265–277. [CrossRef]
23. Hagedoorn, J.G. A process of seismic reflection interpretation. *Geophys. Prospect.* **1954**, *2*, 85–127. [CrossRef]
24. Schneider, W.A. Integral Formulation for Migration in Two and Three Dimensions. *Geophysics* **1978**, *43*, 49–76. [CrossRef]
25. Yilmaz, Ö. *Seismic Data Analysis*, 2nd ed.; Society of Exploration Geophysicists: Tulsa, OK, USA, 2001; Volume 1.
26. Zhuge, X.; Savelyev, T.; Yarovoy, A.G.; Ligthart, L. UWB array-based radar imaging using modified Kirchhoff migration. In Proceedings of the 2008 IEEE International Conference on Ultra-Wideband, Hannover, Germany, 10–12 September 2008; Volume 3, pp. 175–178.
27. Phong, B.T. Illumination for computer generated pictures. *Commun. ACM* **1975**, *18*, 311–317. [CrossRef]
28. Ellis, D.D.; Crowe, D.V. Bistatic reverberation calculations using a three-dimensional scattering function. *J. Acoust. Soc. Am.* **1991**, *89*, 2207–2214. [CrossRef]
29. Hogbom, J. Aperture synthesis with a non-regular distribution of interferometer baselines. *Astron. Astrophys. Suppl.* **1974**, *15*, 49.
30. Charous, A.; Lermusiaux, P.F.J. Dynamically Orthogonal Runge–Kutta Schemes with Perturbative Retractions for the Dynamical Low-Rank Approximation. *SIAM J. Sci. Comput.* 2022, *in press*.
31. Bhabra, M.S.; Lermusiaux, P.F.J. *Stochastic Dynamically Orthogonal Acoustic Wavefront Equations*; Massachusetts Institute of Technology: Cambridge, MA, USA, 2023; Manuscript in preparation.
32. Haley, P.J., Jr.; Lermusiaux, P.F.J. Multiscale two-way embedding schemes for free-surface primitive equations in the "Multidisciplinary Simulation, Estimation and Assimilation System". *Ocean. Dyn.* **2010**, *60*, 1497–1537. [CrossRef]
33. Haley, P.J., Jr.; Agarwal, A.; Lermusiaux, P.F.J. Optimizing Velocities and Transports for Complex Coastal Regions and Archipelagos. *Ocean. Model.* **2015**, *89*, 1–28. [CrossRef]
34. Lu, P.; Lermusiaux, P.F.J. Bayesian Learning of Stochastic Dynamical Models. *Phys. Nonlinear Phenom.* **2021**, *427*, 133003. [CrossRef]
35. Subramani, D.; Lermusiaux, P.F.J. *Probabilistic Ocean Predictions with Dynamically-Orthogonal Primitive Equations*; Massachusetts Institute of Technology: Cambridge, MA, USA, 2023; Manuscript in preparation.
36. Ali, W.H.; Bhabra, M.S.; Lermusiaux, P.F.J.; March, A.; Edwards, J.R.; Rimpau, K.; Ryu, P. Stochastic Oceanographic-Acoustic Prediction and Bayesian Inversion for Wide Area Ocean Floor Mapping. In Proceedings of the OCEANS 2019 MTS/IEEE SEATTLE, Seattle, WA, USA, 27–31 October 2019; pp. 1–10. [CrossRef]
37. Ali, W.H.; Lermusiaux, P.F.J. Dynamically Orthogonal Narrow-Angle Parabolic Equations for Stochastic Underwater Sound Propagation. *J. Acoust. Soc. Am.* 2023, *manuscript under review*.
38. Charous, A.; Lermusiaux, P.F.J. Dynamically Orthogonal Differential Equations for Stochastic and Deterministic Reduced-Order Modeling of Ocean Acoustic Wave Propagation. In Proceedings of the OCEANS 2021, San Diego, CA, USA, 20–23 September 2021; pp. 1–7. [CrossRef]

39. Ali, W.H.; Lermusiaux, P.F.J. *Dynamically Orthogonal Wide-Angle Parabolic Equations for Stochastic Underwater Sound Propagation*; Massachusetts Institute of Technology: Cambridge, MA, USA, 2023; Manuscript in preparation.
40. Ali, W.H.; Lermusiaux, P.F.J. *Joint Ocean-Acoustic Inversion Using Gaussian Mixture Models and the Dynamically Orthogonal Parabolic Equations*; Massachusetts Institute of Technology: Cambridge, MA, USA, 2023; manuscript in preparation.
41. Gupta, A.; Lermusiaux, P.F.J. Neural Closure Models for Dynamical Systems. *Proc. R. Soc.* **2021**, *477*, 1–29. [CrossRef]
42. Gupta, A.; Lermusiaux, P.F.J. *Bayesian Learning of Coupled Biogeochemical-Physical Models*; Massachusetts Institute of Technology: Cambridge, MA, USA, 2023; manuscript under review. Available online: https://arxiv.org/abs/2211.06714 (accessed on 6 March 2023).
43. Lolla, T.; Haley, P.J., Jr.; Lermusiaux, P.F.J. Path planning in multiscale ocean flows: Coordination and dynamic obstacles. *Ocean. Model.* **2015**, *94*, 46–66. [CrossRef]
44. Kulkarni, C.S.; Lermusiaux, P.F.J. Three-dimensional Time-Optimal Path Planning in the Ocean. *Ocean. Model.* **2020**, *152*, 101644. [CrossRef]
45. van Cittert, P.H. Die wahrscheinliche Schwingungsverteilung in einer von einer Lichtquelle direkt oder mittels einer Linse beleuchteten Ebene. *Physica* **1934**, *1*, 201–210. [CrossRef]
46. Zernike, F. The Concept of Degree of Coherence and Its Application to Optical Problems. *Physica* **1938**, *5*, 785–795. [CrossRef]

geomatics

MDPI

Article

Performing a Sonar Acceptance Test of the Kongsberg EM712 Using Open-Source Software: A Case Study of Kluster

Eric Younkin * and S. Harper Umfress *

NOAA National Ocean Service, Office of Coast Survey, Hydrographic Systems and Technology Branch, Silver Spring, MD 20910, USA
* Correspondence: eric.g.younkin@noaa.gov (E.Y.); samuel.umfress@noaa.gov (S.H.U.)

Abstract: In the world of seafloor mapping, the ability to explore and experiment with a dataset in its raw and processed forms is critical. Kluster is an open-source multibeam data processing software package written in Python that enables this exploration. Kluster provides a suite of multibeam processing features, including analysis, visualization, gridding, and data cleaning. We demonstrated these features using a recently acquired dataset from a Kongsberg EM712 multibeam echosounder aboard NOAA Ship *Fairweather*. This test dataset served to illustrate the fundamental analysis abilities of the software, as well as its utility as a troubleshooting tool both in the field and during post-processing. Kluster has the capability to perform the Sonar Acceptance Test in full, including common experiments like the patch test, extinction test, and accuracy test, which are generally performed on new systems. When questions arise regarding the integration or parameter settings of a system, this software allows the user to quickly and clearly visualize much of the raw data and its associated metadata, which is a vital step in any investigative effort. With its emphasis on accessibility and ease of use, Kluster is an excellent tool for users who are inexperienced with multibeam sonar data processing.

Keywords: hydrography; multibeam; open-source; python; Kluster; hydrographic software

Citation: Younkin, E.; Umfress, S.H. Performing a Sonar Acceptance Test of the Kongsberg EM712 Using Open-Source Software: A Case Study of Kluster. *Geomatics* **2022**, *2*, 540–553. https://doi.org/10.3390/geomatics2040029

Academic Editors: Giuseppe Masetti, Ian Church, Anand Hiroji and Ove Andersen

Received: 7 October 2022
Accepted: 25 November 2022
Published: 29 November 2022

1. Introduction

The wealth of resources available for scientific processing and analysis in Python is growing every day. This includes the extensive library of algorithms found in SciPy [1], powerful n-dimensional data structures in NumPy [2], and, more recently, the Pangeo ecosystem [3], which includes packages such as Dask, Xarray, and Zarr. All these packages allow for rapid prototyping of applications and detailed analysis without the required effort to implement existing algorithms and data structures. If processed multibeam data were available in Python, these packages could be used by the scientific community to access the data in a way that is not currently available.

Kluster [4–6] is designed to thrive in this space. It relies on Zarr and Numpy for n-dimensional data structures in memory and on disk. It uses Xarray and Dask to support multiprocessing across all of its processing algorithms. With the core structure being the Xarray Dataset, scientists can read and operate on processed Kluster datasets without using Kluster, relying solely on the Xarray package. Using Zarr, the Kluster datasets are pre-chunked for efficient access over the internet, making the Kluster format an efficient archival format.

Several multibeam processing packages have already been developed and implemented in the open-source space. Most notably, MB-System [7], originally developed at the Lamont-Doherty Earth Observatory of Columbia University using the C language, and SonarScope [8], developed at Ifremer. MB-System is widely considered to be the best alternative to commercial software but has a steep learning curve and requires an understanding of scripting in Linux and command line usage [9]. SonarScope is developed in MATLAB, and is available on Ifremer's GitLab repository, and provides a more complete graphically driven experience. While C and MATLAB are well established languages for development, Python is generally more accessible and an easier language to work in within the scientific community, where many individuals might not have a computer science degree.

To demonstrate the features and effectiveness of Kluster, this paper will outline the Sonar Acceptance Test (SAT) for the newly purchased Kongsberg EM712 multibeam echosounder (MBES) conducted on the NOAA Ship *Fairweather* in May and June of 2022. The SAT encompasses all integration, data acquisition, and processing that is required to ensure that the new sonar meets charting specifications [10]. This is generally a manual process, completed with a combination of Python scripts, commercial software, and minor software development when required. With Kluster, all SAT tests are integrated into the graphical interface, making them simple to run and visualize. The SAT provides a comprehensive test for a multibeam data processing software package, as there are often data issues that interfere with processing, and many custom needs for analysis outside of the standard workflow. As is outlined below, the unique capabilities of Kluster were leveraged throughout this project and proved vital in the qualification of the sonar system under evaluation.

2. Materials and Methods

All software developed in the Kluster project is available on GitHub for download and user contribution. Kluster relies on several custom submodules, that are also available on GitHub in separate repositories, shown in Table 1. Kluster provides build instructions in the documentation, as well as Windows builds for each new release.

Table 1. Kluster and submodule URLs.

Module	URL
Kluster	https://github.com/noaa-ocs-hydrography/kluster (accessed on 22 September 2022)
bathycube	https://github.com/noaa-ocs-hydrography/bathycube (accessed on 22 September 2022)
vyperdatum	https://github.com/noaa-ocs-hydrography/vyperdatum (accessed on 22 September 2022)
drivers	https://github.com/noaa-ocs-hydrography/drivers (accessed on 22 September 2022)
bathygrid	https://github.com/noaa-ocs-hydrography/bathygrid (accessed on 22 September 2022)

Kluster currently supports the Kongsberg .all and .kmall formats, with additional limited support for EK60 and EK80 systems, including a custom amplitude detection capability. Kluster also supports the Reson .s7k format, as is detailed in the Kluster documentation [5] section on 'Requirements'.

The raw dataset for the EM712 SAT is not currently available online, due to limitations with hosting large datasets that are not a part of the normal production chain.

2.1. *Kluster—Theory of Operation*

Kluster first relies on a conversion step, to pull records from the raw multibeam format to an intermediate custom format that was designed for Kluster. This format is stored on disk as Zarr arrays and can be loaded in Xarray as a Dataset, sorted by time and beam. Conversion will automatically sort incoming data into containers, where each container is a specific sonar model, date and sonar serial number. As an example, the extinction test for this experiment exists within container "em712_10070_05_10_2022", which is the container for the EM712 with serial number 10070 on 10 May 2022. Having this organization allows the user to drag-and-drop files into Kluster without concern for which day or sonar they originate from, information that is oftentimes not clear to the end user that was not involved during acquisition.

The initial stages of processing in Kluster are heavily inspired by existing academic research on post-processing multi-sector multibeam systems [11]. Using vessel attitude and mounting angle offsets for the sonar, Kluster corrects the original array-relative beam angles and saves the corrected angle and azimuth to disk, as illustrated in Figure 1. This process currently assumes the transmit and receive arrays are concentric, which may create issues in deeper water, and is a current area of academic interest [12,13]. These values are used during sound velocity correction to calculate the correct offsets from reference point to beam end point, which is then used during georeferencing to build the three-dimensional point cloud. The products of these processing steps are then used in Kluster's Total Propagated Uncertainty (TPU) model, which was built following guidance from the paper on the multibeam uncertainty model [14].

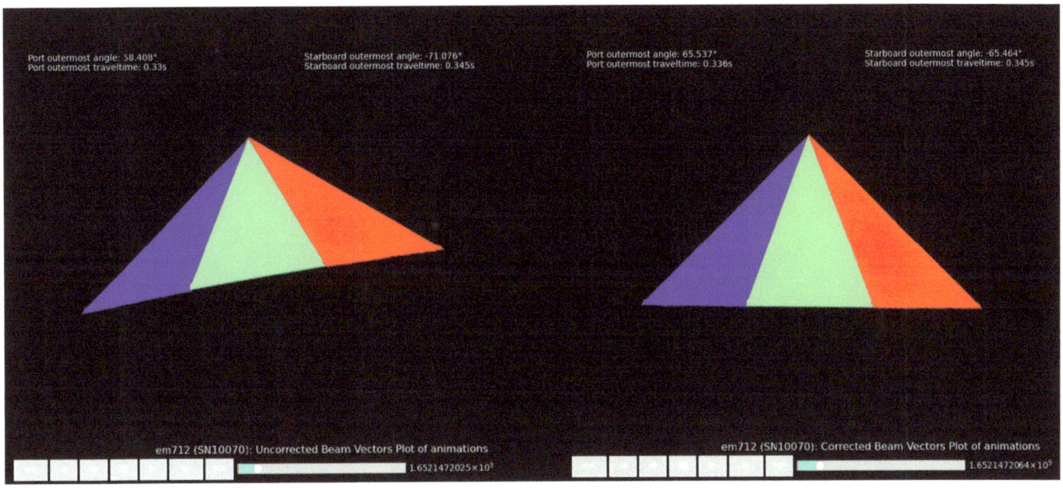

(**a**) (**b**)

Figure 1. Kluster animations of uncorrected and corrected beam vectors available in the Basic Plots tool. Illustrates: (**a**) Raw beam angles as seen in the multibeam data format and; (**b**) Raw angles corrected for attitude and mounting angles as a result of Kluster processing. Colored by multibeam sector.

The user may elect to visualize or manually remove any outliers using the Points View, which displays the point cloud in two or three dimensions. Kluster also provides a filtering utility with some custom filters provided, as well as support for custom filter plugins that can be created by the user for their specific needs.

With a processed point cloud, the user can generate grids using the bathygrid module, which builds single or variable resolution tiles, again saved to disk using Zarr and Xarray. These grids support larger-than-memory datasets, store both points and cell values, support updates through adding and removing additional datasets, and allow for exporting to common GDAL formats.

2.2. NOAA Ship Fairweather & the Kongsberg EM712—Background

NOAA Ship *Fairweather* is 231-foot long hydrographic survey vessel homeported in Ketchikan, Alaska (Figure 2). Commissioned in 1968, the ship operates an EM712 sonar and carries a variety of small boats with individual sonars and additional charting capabilities. The Kongsberg EM712 installed on NOAA Ship *Fairweather* is a 0.5° × 1.0° system with a specified maximum depth of 3200 m and a maximum coverage of 3950 m. This system is controlled by the latest version of Kongsberg's SIS5 software and is one of the first instances of this software in the NOAA fleet. There were several integration issues with this software that were resolved prior to sailing, mostly centered around interfacing with other software packages. The EM712 receives attitude, velocity, and navigation from the Applanix POS MV installed on the vessel. The EM712 transmitter serves as the vessel reference point, and all offsets and angles are relative to it. These offsets and angles are entered into SIS and the POS MV setup screens such that the raw multibeam data is logged with all the correct values.

Figure 2. Image of NOAA Ship *Fairweather*, courtesy of NOAA.

2.3. Ancillary Data Processing

During this project, sound velocity profiles were acquired using a Moving Vessel Profiler (MVP) winch system and an AML Micro CTD sensor. These raw profiles are processed in Sound Speed Manager, which is an open-source sound velocity processing software available through Hydroffice [15]. Kluster supports importing these processed sound velocity text files as additional profiles to those currently in the raw multibeam data. Sound velocity profiles are used during sound velocity correction in Kluster based on one of the available selection algorithms seen in the Kluster project settings.

Raw POS MV data is processed in Applanix POSPac using the Trimble RTX corrector service to produce processed GNSS/INS data in the Applanix Smoothed Best Estimate of Trajectory (SBET) format. Kluster can import the processed navigation and ellipsoid height for use in all georeferencing operations, which is of particular significance with ellipsoidally referenced surveying (ERS) techniques where the final depth is a product of the ellipsoid height [16]. Processed navigation is generally used throughout all vertical datum selections in Kluster.

3. Results

The EM712 SAT took place off the coast of San Francisco, California, USA. All data were converted in Kluster by simply dragging in the raw Kongsberg KMALL files and using the start button in the Action pane to commence conversion. These files are shown in the screenshot below in Figure 3 with the project information on the bar on the left, and the tracklines shown in the embedded QGIS map view on top of an OpenStreetMap WMS layer. By utilizing QGIS tools for the map view in Kluster, resources such as web map services and generic raster and vector format support are made available in a simple and intuitive way.

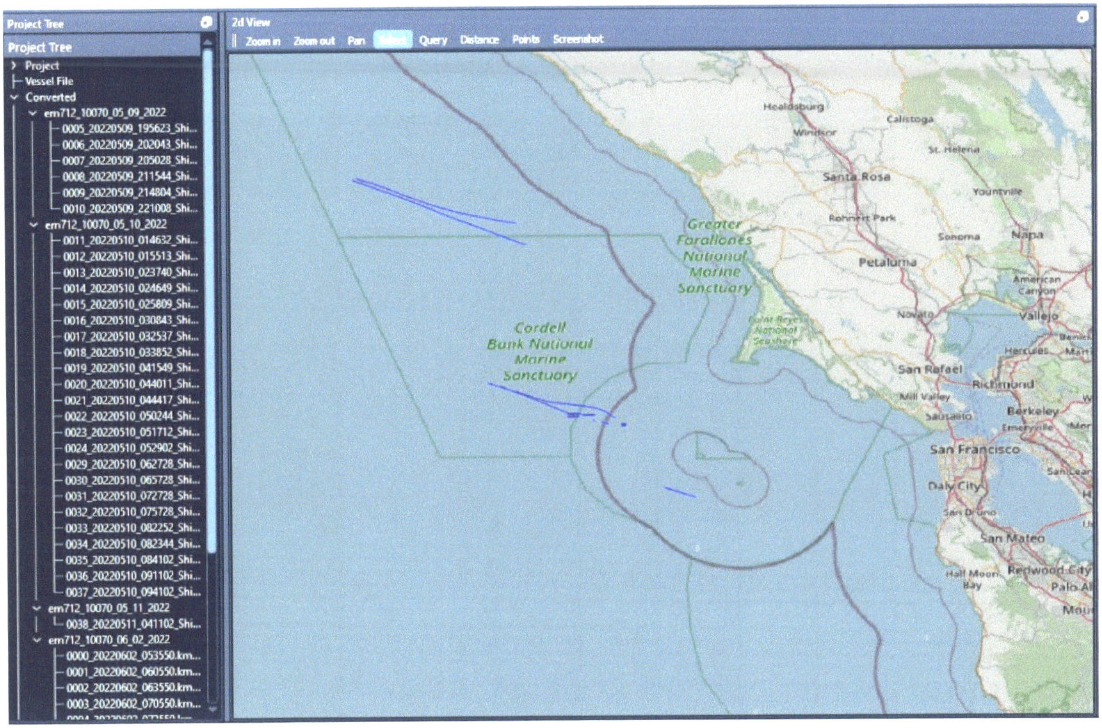

Figure 3. Map view of the project area in Kluster, multibeam tracks shown in blue.

Kluster has a custom state, machine-driven processing system called the Intelligence module, which generates processing actions based on the state of the data and the desired processing settings. Dragging in new multibeam files that do not exist in the existing containers will generate a new conversion action. Including additional sound velocity profiles will generate a new import sound velocity action. If sound velocity processing had occurred once already, importing new profiles would generate a re-sound velocity correct action. The Intelligence module will ensure that the data is fully processed as project settings change and new data is added. For this dataset, we processed to the NOAA mean lower low water (MLLW) datum, using the processed ellipsoid height from the SBET and the vyperdatum module to automatically generate a separation model between ellipsoid and MLLW, as shown in Figure 4.

With the newly processed data, we are now able to proceed to the SAT tests. NOAA's SAT procedure generally includes the following tests, which will dictate the layout of the rest of this section:

- Offsets and Integration
- Patch Test (Boresight Angle Estimation)
- Extinction Test (Range Test)
- Accuracy Test (Vertical Accuracy Test)

3.1. Offsets and Integration

Kluster includes on conversion all offsets and supporting parameters that SIS can provide in the KMALL file. These are shown in the container attribution in the Attribute window but are primarily interacted with through the Vessel Setup utility. The Vessel Setup utility allows the user to see and change the offsets and setup parameters within the container selected. Additionally included is a few 3D models of ships that can be used as a reference for the blocks that represent the sensor locations. The user can also include a 3D model for their vessel to visualize the sensor locations in OBJ format. Figure 5 shows the sonar transmitter and receiver for this survey, with the offsets from the vessel reference point on the left.

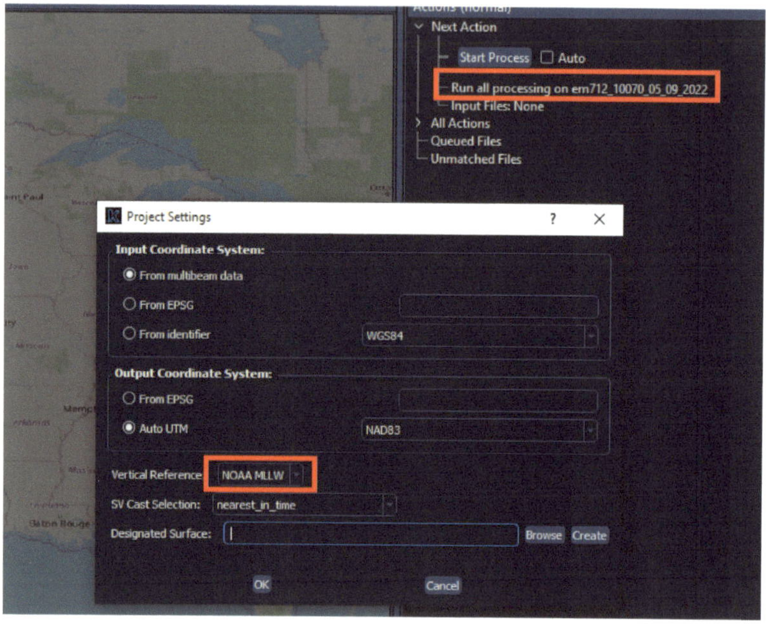

Figure 4. Project settings with desired vertical reference and the "Run all processing" action that is spawned as a result.

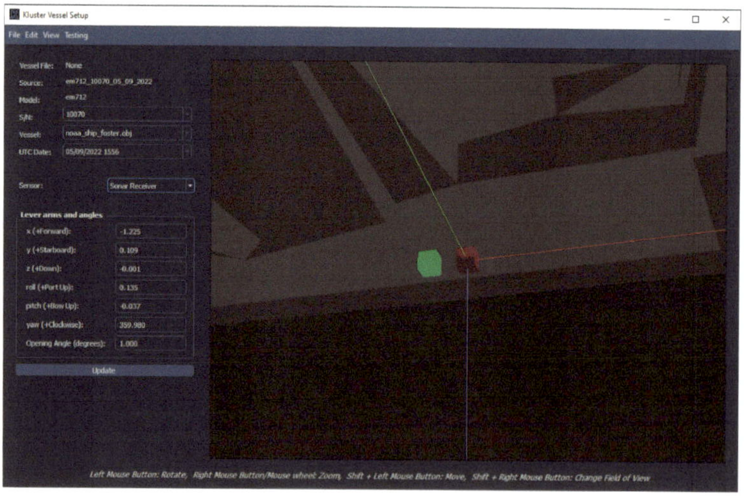

Figure 5. Vessel Setup utility showing the sonar transmitter (red) and receiver (green).

Changes to these parameters will spawn the appropriate processing action, depending on the value altered. Mounting angle changes require a full reprocessing of the dataset, while changes to uncertainty parameters will only spawn an uncertainty processing step. These values in the Vessel Setup match the transmitter-relative values entered into SIS, so no further action is required.

3.2. Patch Test

The patch test, or boresight angle estimation, includes six survey lines run on 9 May 2022, that are processed and evaluated to determine any residual angular offsets or timing offset between the sonar transducer and the motion sensing unit. These latency, pitch, roll, and yaw offsets are determined by comparing the bathymetry of lines collected at different orientations over both flat and sloped seafloor. As a result, we initially needed to generate a processed grid to assess the acquired bathymetry. Figure 6 illustrates this, showing an 8.0 m resolution Combined Uncertainty and Bathymetry Estimator (CUBE) [17] grid, with data processed to the NOAA MLLW ERS datum. CUBE is provided in the bathycube module that was developed alongside Kluster. Depths range from 100 m to 700 m.

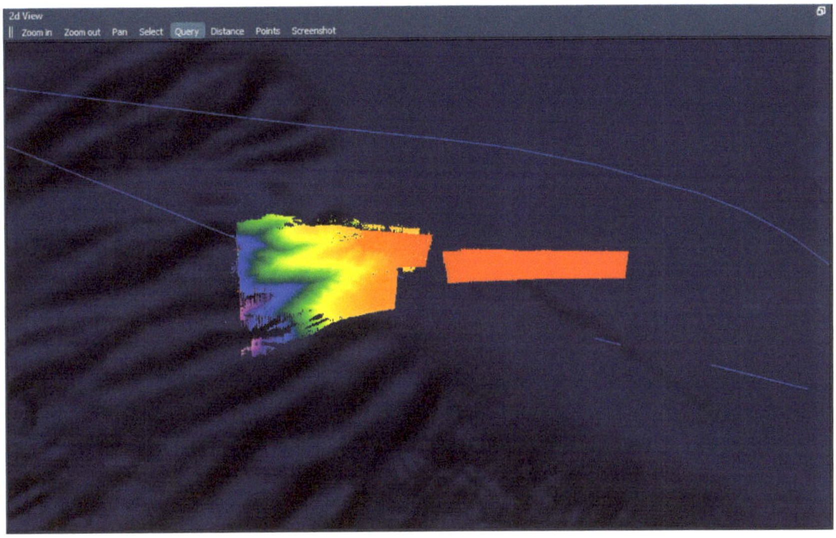

Figure 6. Patch Test Area with an 8.0 m resolution gridded dataset of the processed multibeam data.

The latency test involves isolating a line over flat seafloor that was acquired with, ideally, a significant amount of roll. This dataset can be analyzed using the Kluster Advanced Plots—Wobble Test tool to determine if there is a correlation between the roll rate and the ping slope, where the slope of the regression would be the calculated latency between sonar and motion sensor. In the case of line 0010, we were unable to determine any significant latency value. If we had computed a value of several milliseconds or greater, we would enter it into SIS, before commencing any other tests. Alternatively, the value can be added in Kluster when post-processing the dataset. Figure 7 illustrates the Latency Test as completed in Kluster.

The remaining three elements of the Patch Test can be determined using the Kluster Patch Test tool. Roll is calculated using the same line run twice in opposite directions over a flat seafloor, pitch is calculated using the same line run twice in opposite directions over a slope, and yaw involves two lines run in the same direction down a slope offset from each other. These lines can be chosen from the six included in the Patch Test dataset, which were specifically acquired to meet these guidelines. To accomplish the Patch Test, new values are chosen by the user and entered into the utility, to reprocess the data displayed in the point cloud viewer until the data is visually determined to be in alignment and acceptable. Figure 8 shows this process, with a narrow slice of the dataset perpendicular to the vessel motion shown in the points view, colored by the line of origin.

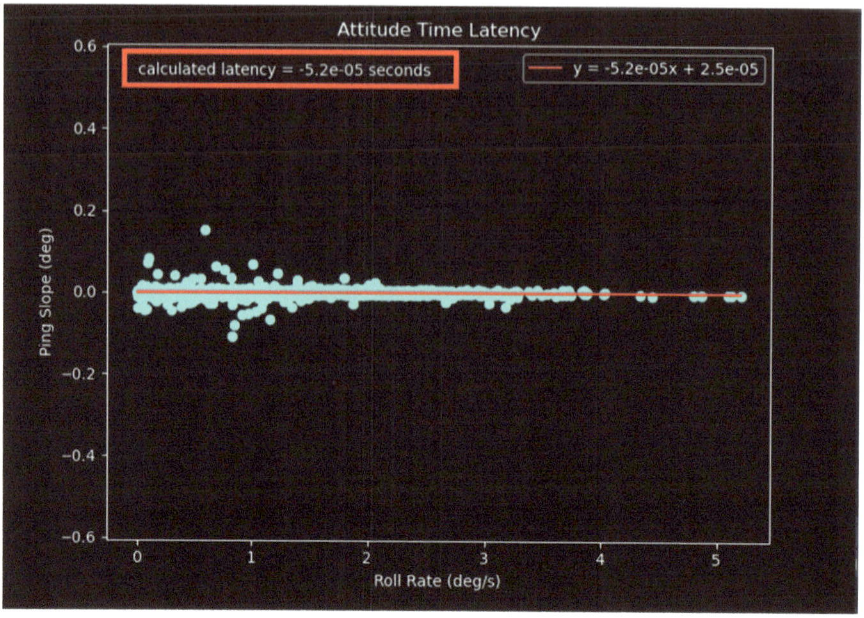

Figure 7. Latency Test shown in the Advanced Plots tool.

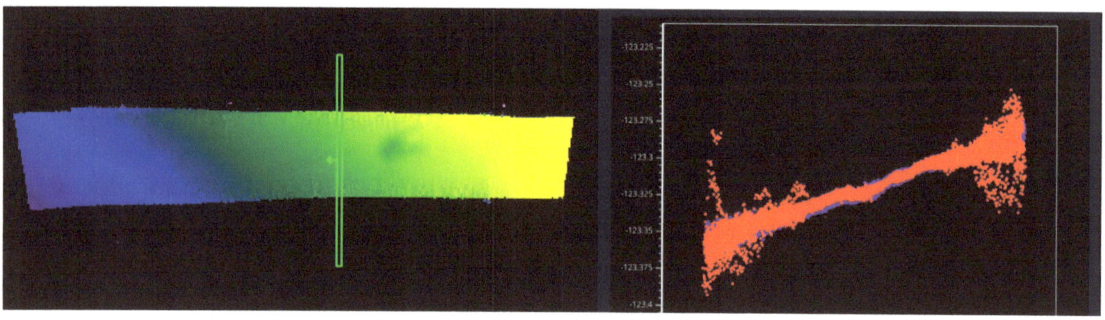

Figure 8. Patch Test utility shown, with data being assessed for roll mounting angle offset.

If new values are found, they can be entered into the Vessel Setup utility for reprocessing the existing dataset; though ultimately, they should be entered in SIS, such that the raw data will already have the correct mounting angles.

3.3. Extinction Test

The extinction test serves to determine the effective swath width of the system throughout the expected depth range, as well as the system's ability to automatically select the appropriate depth-dependent settings. The survey lines are generally run from shallow to deep and then deep to shallow following the reciprocal course. The resulting data can be plotted using the Kluster Advanced Plots tool to visualize the outermost beams seen for each depth range. The extinction test area and acquired bathymetry are shown in Figure 9 below. After the completion of one extinction test, we noted non-uniform changes in the depth settings and underperformance of the sonar. We used Kluster to diagnose the issue and subsequently reacquired the dataset.

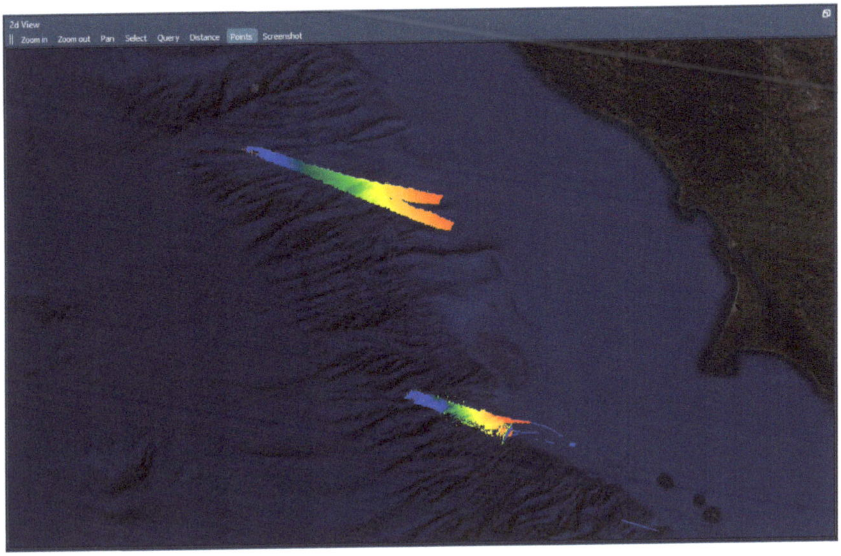

Figure 9. Extinction test area and gridded datasets with 64 m resolution shown on top of the Satellite WMS layer.

The extinction test results are a series of plots showing the width of the swath at different depths and colored by sonar mode setting or frequency. The first extinction test results are shown in Figure 10 below. Each of the three plots shows the progression from shallow settings to deep settings. In the case of frequency, it goes from high to low with increasing depth; for ping mode, we see Frequency Modulated (FM) mode engaged toward the deeper range of the system; for depth mode, we see the system step from very shallow up to very deep mode. While these trends are generally expected, these particular plots also display a curious lack of uniformity in their progression, namely with regard to the switching of mode two shown in Figure 10c. When compared against the associated bathymetry, these extinction lines hinted at either a malfunctioning sonar or a misconfiguration of the operating parameters.

This prompted an examination of the sonar settings to determine if anything was amiss. We determined that the Angular Coverage mode was set to Manual, fixing the swath angle instead of allowing it to dynamically adjust based on the operating conditions and depth. This setting, as well as all other runtime settings, can be seen in the Kluster Attribution window, shown in Figure 11.

With the manual Angular Coverage mode limiting the performance of the sonar and resulting in poor outer beam performance, a second test was planned to determine the appropriate swath width relationship. This test is shown below in Figure 12. The mode change resulted in a much cleaner swath as the system compensated for depth by adjusting the beam angles appropriately. The system did not perform as well as expected, both in terms of ultimate depth range and swath width in deeper waters, but due to the heavy weather seen during this test, it was not entirely unexpected. Additionally, in the second extinction test, the minimum frequency was intentionally set to 70 kHz, instead of 50 khz, resulting in a narrower swath in deeper depths, as compared to the first test. This is reflected in Figures 10a and 12a.

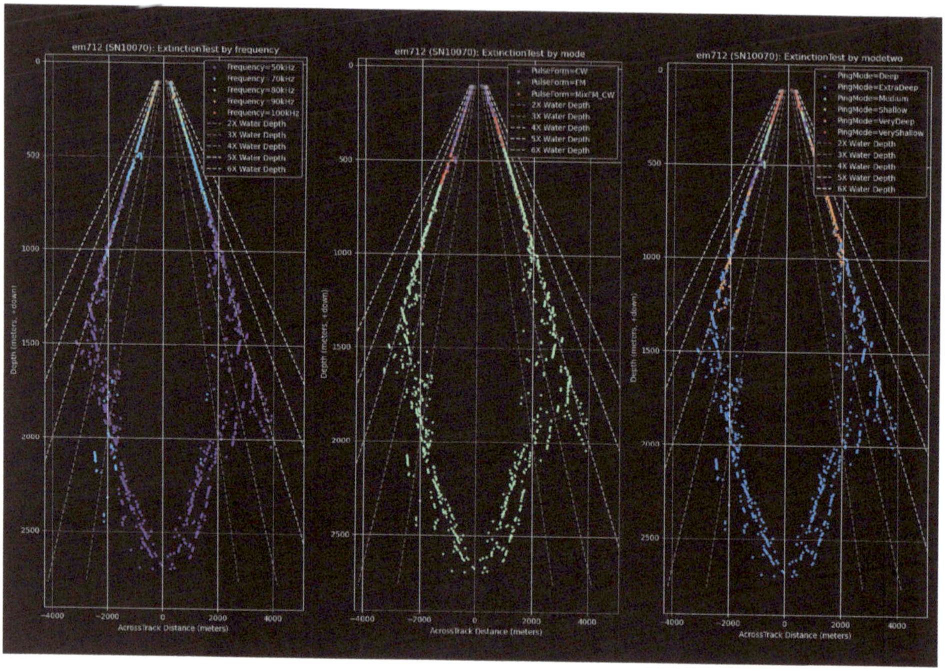

(a) (b) (c)

Figure 10. Kluster Extinction Test results for the initial test. Illustrates: (**a**) Swath width versus depth colored by frequency; (**b**) Swath width versus depth colored by the first mode value, which is either a Continuous Wave (CW) pulse or a Frequency Modulated (FM) pulse; (**c**) Swath width versus depth colored by the second mode value, which is the Kongsberg Ping Mode.

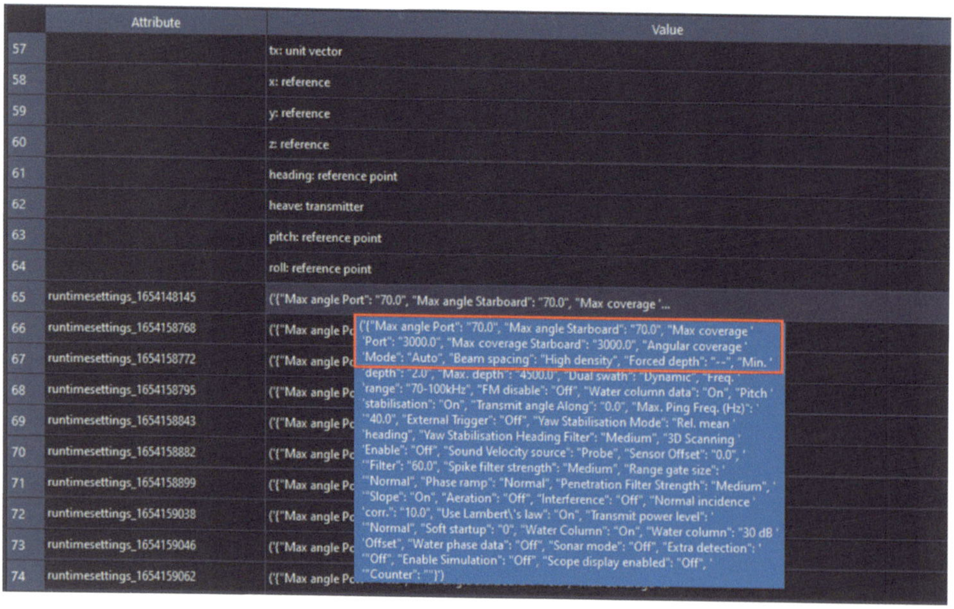

Figure 11. Runtime parameters as shown in the Kluster Attribute window. Illustrates the runtime parameters for the second extinction test, where the Angular Coverage was changed to Auto.

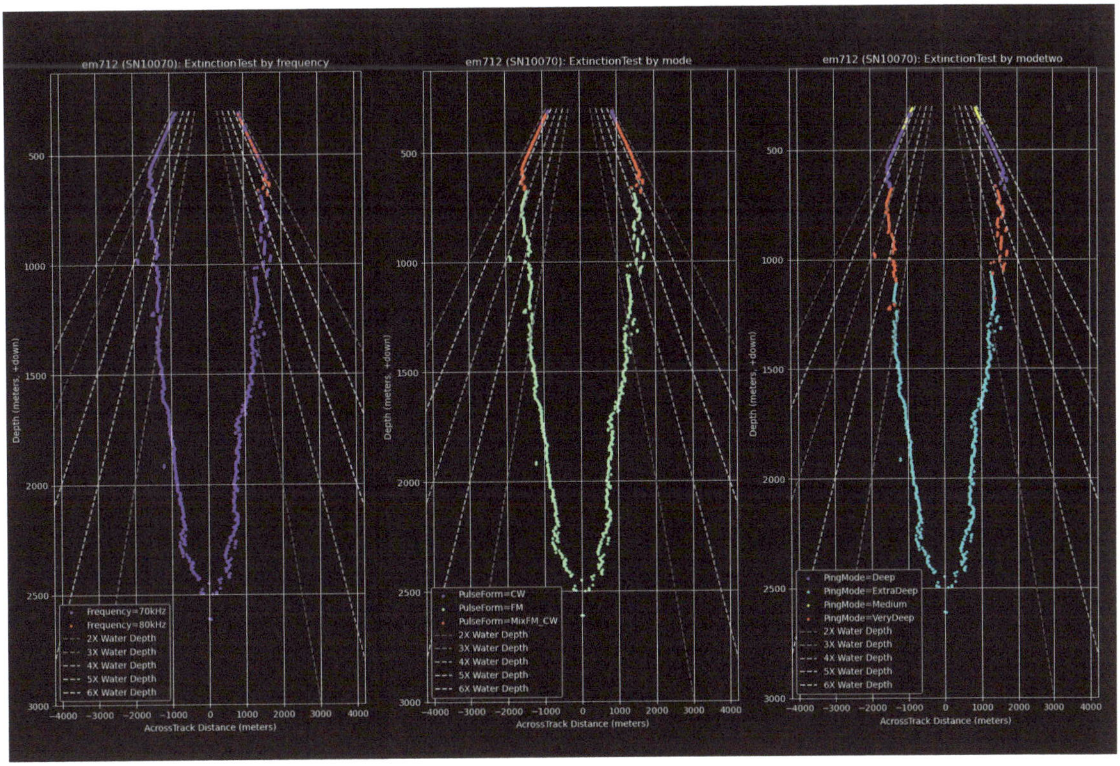

(a) (b) (c)

Figure 12. Kluster Extinction Test results for the second test. Illustrates: (**a**) Swath width versus depth colored by frequency; (**b**) Swath width versus depth colored by the first mode value, which is either a Continuous Wave (CW) pulse or a Frequency Modulated (FM) pulse; (**c**) Swath width versus depth colored by the second mode value, which is the Kongsberg Ping Mode.

3.4. Accuracy Test

The accuracy test is a means of determining the internal consistency of the sonar's seafloor measurements as a function of the beam angle, looking specifically for vertical differences between reference and acquired data. This is accomplished by comparing test lines against a densely populated, gridded dataset. First, we drove parallel lines to acquire a high density dataset over a relatively flat seafloor area. This dataset populated a reference bathymetry grid, against which we compared our accuracy lines. The accuracy lines were driven orthogonal to the set of reference tracklines, in pairs associated with each frequency and depth mode. In this way, the accuracy test allows the sonar operator to identify and isolate areas of high depth uncertainty and determine if they are a function of mode, frequency, or beam-angle. NOAA Ship *Fairweather* collected accuracy test data in multiple modes, frequencies, and depth regimes. In this paper, we focus on the accuracy test results gathered in roughly 250 m of water, in medium depth mode.

After processing the raw mutlibeam data, we imported post-processed navigation which automatically initiated a new cycle of georeferencing for the lines. Processing the accuracy test itself in Kluster occured in two independent steps. First, we selected the lines associated with the reference grid and created a surface. In this case, we created a variable resolution grid with depths computed by the CUBE algorithm. We then selected the accuracy test lines and used the Advanced Plots tool to conduct a grid-to-sounding comparison and output our beam-wise and angle-wise comparison plots. Typically, accuracy test lines are run in succession as the sonar operator shifts through the various frequencies and modes. Kluster automatically groups these lines by frequency and mode before conducting the comparison. In this example, we selected a total of four lines, two in 70–100 kHz medium mode and two in 70–100 kHz deep mode. Kluster outputs accuracy plots corresponding to each pair respectively. The depth bias plot as a function of beam angle for 70–100 kHz in medium depth mode is shown below in Figure 13.

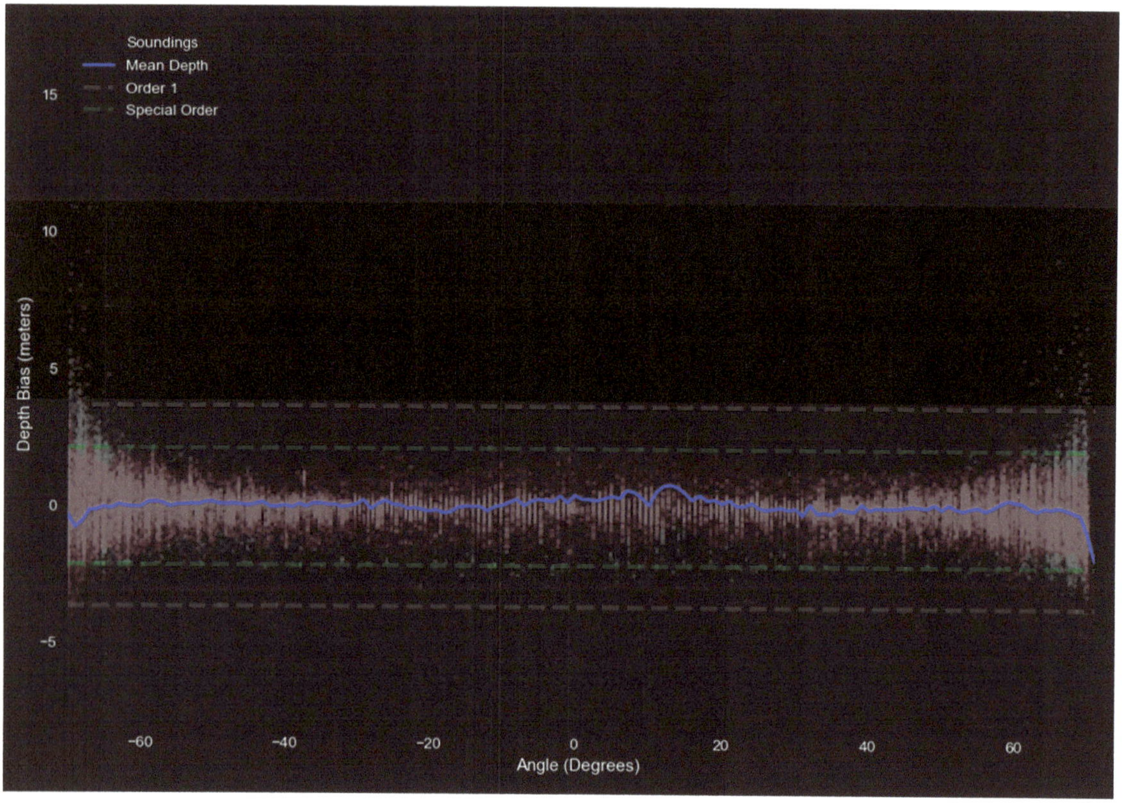

Figure 13. Kluster Accuracy Test result, showing the depth bias between the accuracy test lines in 70–100 kHz, medium mode, and the reference grid, plotted as function of beam angle. Comparisons to IHO Order 1 and IHO Special Order are shown as the horizontal dotted lines.

When plotting the grid-to-sounding comparison, Kluster automatically computes the average depth offset between the accuracy soundings and the reference surface. This bias is then removed from the computed result and displayed at the top of the chart, which enables a more coherent visualization of the small differences in depth bias as a function of angle or beam. In this example, the average bias was −3.5 cm, meaning the soundings were, on average, 3.5 cm shallower than the reference grid. The additional plotting of the Order 1 and Special Order specification for the reference surface depth allows us to quickly confirm that this sonar meets the requirement for uncertainty in the displayed mode.

4. Discussion

Kluster provides a new and intuitive way to process and analyze multibeam data. Through the integrated intelligence module, Kluster will spawn the appropriate actions for the user to take. This eliminates the need for the user to intimately understand the idiosyncrasies of multibeam processing, as is commonly required by other existing software. Experienced hydrographers and inexperienced users seeking to access multibeam data alike stand to benefit from this simplified workflow. Through the Sonar Acceptance Test toolset, Kluster also supports analysis of the state of the sonar itself, in a way not currently available in other open-source software.

Kluster currently includes support for only Kongsberg (.all, .kmall, .raw) and Reson (.s7k) formats, but with the intermediate Xarray/Zarr format that Kluster generates on data conversion, the system has the capacity to support other sonar systems in the future. This future work item is a high priority of the project, as it directly enables the growth of the community around the software package. Kluster is available on a publicly accessible GitHub repository [6]. New releases include Windows builds of the software package for users unfamiliar with the creation of a Python environment.

Being entirely written in Python, Kluster is an attractive project for developers of all skill levels, as interacting and building off Kluster is a relatively simple matter. With the development of plugins, such as the Filter Module in Kluster, community engagement with Kluster can be made even easier—supporting experimentation in multibeam processing in a new and exciting way. Kluster represents a valuable tool for universities, companies, governments, or even individuals that seek to process multibeam data; be it for education, crowd sourced bathymetry, charting acquisition, or any other application of seafloor data.

Author Contributions: Conceptualization, E.Y.; methodology, E.Y.; software, E.Y.; validation, E.Y.; formal analysis, E.Y. and S.H.U.; investigation, E.Y. and S.H.U.; resources, E.Y.; data curation, E.Y. and S.H.U.; writing—original draft preparation, E.Y. and S.H.U.; writing—review and editing, E.Y. and S.H.U.; visualization, E.Y.; supervision, E.Y.; project administration, E.Y. All authors have read and agreed to the published version of the manuscript.

Funding: This research received no external funding.

Institutional Review Board Statement: Not applicable.

Informed Consent Statement: Not applicable.

Data Availability Statement: Access to Kluster is detailed in Section 2. The EM712 acceptance test dataset is unavailable for public download at this time.

Acknowledgments: We would like to extend our gratitude to the crew of NOAA Ship *Fairweather* for their dedication of time, resources, and energy to the acceptance testing of their EM712 sonar and the usage of Kluster.

Conflicts of Interest: The authors declare no conflict of interest.

References

1. Virtanen, P.; Gommers, R.; Oliphant, T.E.; Haberland, M.; Reddy, T.; Cournapeau, D.; Burovski, E.; Peterson, P.; Weckesser, W.; Bright, J.; et al. SciPy 1.0: Fundamental Algorithms for Scientific Computing in Python. *Nat. Methods* **2020**, *17*, 261–272. [CrossRef]
2. Harris, C.R.; Millman, K.J.; Van Der Walt, S.J.; Gommers, R.; Virtanen, P.; Cournapeau, D.; Wieser, E.; Taylor, J.; Berg, S.; Smith, N.J.; et al. Array programming with NumPy. *Nature* **2020**, *585*, 357–362. [CrossRef] [PubMed]
3. Eynard-Bontemps, G.; Abernathey, R.; Hamman, J.; Ponte, A.; Rath, W. The Pangeo Big Data Ecosystem and its Use at CNES. In Proceedings of the 2019 conference on Big Data from Space (BiDS 19), Munich, Germany, 19 February 2019.
4. Younkin, E. Kluster: Distributed Multibeam Processing System in the Pangeo Ecosystem. In Proceedings of the Oceans 2021: San Diego–Porto, San Diego, CA, USA, 20–23 September 2021; pp. 1–6.
5. Kluster Documentation. Available online: https://kluster.readthedocs.io/en/latest/ (accessed on 23 September 2022).
6. Kluster. Available online: https://github.com/noaa-ocs-hydrography/kluster (accessed on 23 September 2022).
7. MB-System Seafloor Mapping Software. Available online: https://www.mbari.org/products/research-software/mb-system (accessed on 22 September 2022).
8. Augustin, J.; Fezzani, R.; Poncelet, C.; Scalabrin, C. *SonarScope Software*; SEANOE: Plouzané, France, 2022. [CrossRef]
9. Cordero, J.M.; Kastrisios, C. Characterizing free and open-source tools for ocean-mapping. In Proceedings of the 6th Hydrographic Engineering Conference, Lisbon, Portugal, 3 November 2020.
10. International Hydrographic Organization. *Standards for Hydrographic Surveys*; S-44 Edition 6.1.0; International Hydrographic Organization: Monaco, 2020.
11. Beaudoin, J.D.; Hughes Clarke, J.E. Application of Surface Sound Speed Measurements in Post-processing for Multi-Sector Multibeam Echosounders. *Int. Hydrogr. Rev.* **2004**, *5*, 17–32.

12. Hamilton, T.; Beaudoin, J.; Hughes Clarke, J.E. A More Precise Algorithm to Account for Non Concentric Multibeam Array Geometry. In Proceedings of the 2014 Canadian Hydrographic Conference, St. John's, NL, Canada, 14 April 2014.
13. Xianhai, B.; Yang, F.; Ma, Y.; Wu, D.; Zhang, K.; Xu, F. Simplified Calibration Method for Multibeam Footprint Displacements Due to Non-concentric Arrays. *Ocean Eng.* **2020**, *197*, 106862.
14. Hare, R. Depth and Position Error Budgets for Multibeam Echosounding. *Int. Hydrogr. Rev.* **1995**, *LXXII*, 37–69.
15. Gallagher, B.; Masetti, G.; Zhang, C.; Calder, B.R.; Wilson, M.J. Sound Speed Manager: An Open-Source Initiative to Streamline the Hydrographic Data Acquisition Workflow. In Proceedings of the 2017 US Hydro Conference, Galveston, TX, USA, 20 March 2017.
16. Mills, J.; Dodd, D. *Ellipsoidally Referenced Surveying for Hydrography*; International Federation of Surveyors (FIG): Copenhagen, Denmark, 2014; Number 62.
17. Calder, B. Automatic Statistical Processing of Multibeam Echosounder Data. *Int. Hydrogr. Rev.* **2003**, *980*, 53–68.

geomatics

Article

Effective Automated Procedures for Hydrographic Data Review

Giuseppe Masetti [1,2,*], Tyanne Faulkes [3], Matthew Wilson [4] and Julia Wallace [4]

1 Center for Coastal and Ocean Mapping, NOAA-UNH Joint Hydrographic Center, University of New Hampshire, Durham, NH 03824, USA
2 Danish Geodata Agency, Danish Hydrographic Office, 9400 Nørresundby, Denmark
3 NOAA National Ocean Service, Office of Coast Survey, Hydrographic Survey Division, Pacific Hydrographic Branch, Seattle, WA 98115, USA
4 NOAA National Ocean Service, Office of Coast Survey, Hydrographic Survey Division, Atlantic Hydrographic Branch, Norfolk, VA 23510, USA
* Correspondence: giuseppe.masetti@unh.edu

Abstract: Reviewing hydrographic data for nautical charting is still a predominately manual process, performed by experienced analysts and based on directives developed over the years by the hydrographic office of interest. With the primary intent to increase the effectiveness of the review process, a set of automated procedures has been developed over the past few years, translating a significant portion of the NOAA Office of Coast Survey's specifications for hydrographic data review into code (i.e., the HydrOffice applications called QC Tools and CA Tools). When applied to a large number of hydrographic surveys, it has been confirmed that such procedures improve both the quality and timeliness of the review process. Increased confidence in the reviewed data, especially by personnel in training, has also been observed. As such, the combined effect of applying these procedures is a novel holistic approach to hydrographic data review. Given the similarities of review procedures among hydrographic offices, the described approach has generated interest in the ocean mapping community.

Keywords: hydrographic survey; automated procedures; data review; nautical charts

Citation: Masetti, G.; Faulkes, T.; Wilson, M.; Wallace, J. Effective Automated Procedures for Hydrographic Data Review. *Geomatics* **2022**, *2*, 338–354. https://doi.org/10.3390/geomatics2030019

Academic Editor: Naser El-Sheimy

Received: 2 June 2022
Accepted: 22 August 2022
Published: 25 August 2022

1. Introduction

The review of hydrographic data for nautical charting is still a predominately manual process, consisting of tedious and monotonous tasks [1,2]. These tasks typically arise from the application of directives developed over the years—and in continuous evolution—by the hydrographic office in charge of nautical charting products for specific regions. The practical interpretation of such directives requires the intervention of experienced analysts applying monotonous data evaluations, which is, by nature, conducive to inconsistencies and human error [3–5].

However, a portion of these directives can be—or have the potential to become—interpreted algorithmically by providing a quantitative translation (e.g., matching thresholds) of what was the original intention of a given rule [6]. Quite often, the algorithmic translation represents an occasion to clarify and improve the text of the initial directives. By focusing on the automation of the most monotonous actions performed, the review process can become significantly faster and more effective, with more efforts dedicated to handling special cases and less common situations. These changes also have the benefit of increasing reproducibility due to the reduction in human subjectivity.

Bathymetric grids are commonly affected by both *fliers*—anomalous depth values resulting from spurious soundings—and *holidays*—empty grid cells due to insufficient bathymetric information [7–9]. In particular, the detection of fliers of different types (e.g., isolated versus clustered) and the effective distinguishing of them from real bathymetric features is a quite challenging task [10–12]. Survey specifications often require that bathymetric grids be free of fliers and large holidays, fulfill statistical metrics, and meet format and metadata requirements.

With the primary intent to reduce the time between the data collection (i.e., sonar pinging) and the publication of the derived products (the *ping-to-public interval*), this work describes a set of automated procedures derived over the past few years, translating a significant portion of the NOAA Office of Coast Survey's Hydrographic Survey Specifications and Deliverables (HSSD) [13] (which are based on the guidelines published by the International Hydrographic Organization) into code. The mentioned code is made accessible through two free and open applications, called QC Tools and CA Tools, respectively [6,14,15].

This work starts by describing the rationale and the design principles of the procedures for quality control of bathymetric grids, validation of significant features, and evaluation of the survey against the nautical chart to assess chart adequacy. Then, the software implementation of these procedures is described, and several meaningful results of their application are highlighted. Finally, conclusions are presented, along with ideas for future work to improve the user interaction with the algorithms.

2. Rationale and Design Principles

A ping-to-public workflow for hydrographic survey data consists of several steps, each of them requiring some level of human intervention. A new paradigm has been adopted which allows the analyst to focus on parts of the data that require remediation, rather than spreading the effort across the entire dataset equally. Specifically, the tools for quality control of survey products have been incrementally developed in the past decade [16], while the tools to assess chart adequacy are based on the seminal work described in [15].

The automated procedures have been developed through stand-alone tools that are agnostic of the software solution adopted in processing the survey data. This approach was chosen to achieve the significant advantage of having the tools act like independent agents, inspecting survey products, evaluating their quality, and thus, increasing the confidence in the original survey submission. The algorithms have been focused on two typical final products of a hydrographic survey—bathymetric grids and feature files—and identify issues common to these data products based on survey specifications. A key requirement for success has been that the resulting tools are easily customizable to meet new and modified agency-specific requirements.

To ease their adoptability, the tools access the survey data through two open formats popular in the ocean mapping field: the International Hydrographic Organization's S-57 format [17] for vector features, and the Open Navigation Surface's Bathymetry Attributed Grid (BAG) format [18] for gridded bathymetry. To avoid preliminary format transformation steps, closed formats have also been added for manufacturers providing an access library. A concrete example is represented by the CARIS Spatial Archive™ (CSAR) format, accessed using the CARIS' CSAR SDK version 2.3.0 [19]. Furthermore, the support of the NOAA Bathygrid format, recently developed as part of NOAA's Kluster project (a distributed multibeam processing system) [20], is currently in the advanced experimental phase. The addition of other formats is facilitated by other leading companies providing libraries to ease the access to their proprietary data formats.

The code has been organized as Python packages [21]. To encourage community involvement and code contributions, the Python language was selected due to its popularity in the geospatial field. The packages have also been designed to be highly modularized.

2.1. Grid Quality Control

The fliers are often associated with suboptimal data filtering and cleaning, both automatic and manual, of high-density hydrographic surveys such as the ones acquired with multibeam echosounders [1,5,22,23]. A hydrographic data reviewer may identify the presence of such fliers using traditional methods, such as inspection using 2D/3D viewers or evaluation of specific grid metrics, and/or shoal-biased sounding selection [24,25]. However, these methods are inherently error-prone and quite subjective, with the result that several fliers can be easily missed during the hydrographic data review [14]. As such, it is not surprising that in 2015, the NOAA Hydrographic Surveys Division reported that nearly 25% of the surveys received were affected by fliers [26]. Even adopting more than one of the methods mentioned, it is challenging to identify all the fliers that may be present on a grid with several millions of cells [16]. Scanning the grid with automated algorithms that flag potential anomalies not only supports the job of the reviewer, but also builds confidence in the performed manual evaluation. This is especially true in areas with rough seafloor morphology, where small fliers can be easily confused with natural features (Figure 1) [27].

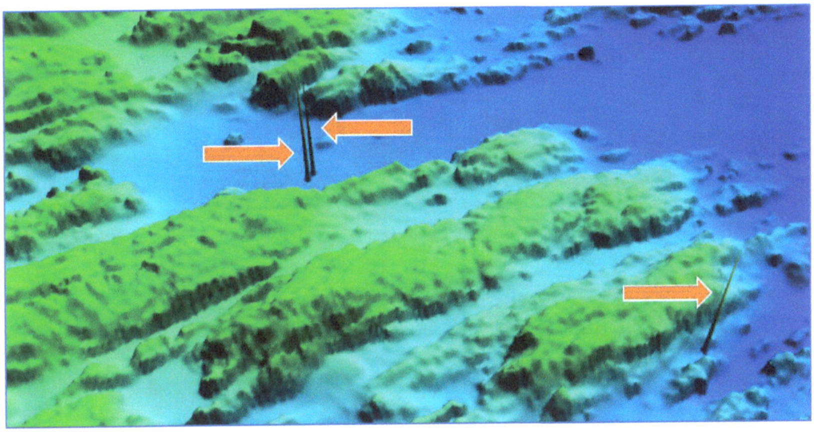

Figure 1. Depth fliers (pointed out by the orange arrows) of a few meters in a bathymetric grid with an average depth of 50 m. Though no algorithm can distinguish them from the natural seafloor with 100% accuracy, human reviewers are aided greatly by automated scanning to *flag* suspect areas. Image obtained using CARIS HIPS and SIPS software.

A manual grid inspection for identification of all the holidays is a comparable challenge [26]. However, while there are different types of fliers (e.g., isolated vs. clustered), the definition of what is considered a significant holiday is quite objective and is usually outlined in the survey requirements [13]. There is great advantage in developing a robust algorithmic translation to automatically scan for potential holidays.

Several hydrographic specifications—for instance, the NOAA Hydrographic Survey Specifications and Deliverables (HSSD) [13]—allow for the manual selection of specific soundings (*designated soundings*) being judged as particularly significant and thus, requiring their depth value to be enforced in the grid. When designated soundings are in use, their automated review is beneficial to evaluate their alignment with the specifications (for instance, to identify the misuse of designated soundings). The alternative to such automated review is tedious, manual work based on vertical or horizontal measurements in the surroundings of each designated sounding.

It is also quite common that the survey specifications have requirements for the grid's specific statistical metrics (e.g., uncertainty, density of soundings) [28]. Although software providers usually support calculation of statistical grid layers, it is not common for the validation against hydrographic specifications to be included. The translation of such rules into an automated procedure—returning a *pass* or *fail* indication and/or providing a visual representation of the rules—has the positive effects of simplifying the job of the reviewer, enforcing consistent interpretation across all the datasets, and making any future customization much easier.

Ensuring that the created products fulfill format specifications (e.g., the BAG Format Specification Document [18]) is also of great value. Such a fulfillment eases the data interoperability, ensuring that internal and public users of a survey bathymetric grid can properly access and interpret the collected survey data.

2.2. Significant Features Validation

The outcome of a hydrographic survey is not usually limited to a bathymetric point cloud and the bathymetric grid derived from it. The surveyor is quite often called to integrate the collected bathymetry with a set of significant features. These features may carry a variety of information that may interest the seafarers, such as dangers, or auxiliary aids to navigation. Although several manufacturer-specific methods for feature validation exist, it is beneficial for a hydrographic office to be able to not only enforce specific feature validation tests, but also to run them independently of the specific processing software in use.

In approach and harbor areas, the number of significant features can be large and the review of the associated metadata time consuming, error-prone, and particularly tedious (Figure 2). In addition, the task at hand is made even more challenging by the necessity of adhering to all the rules required to ensure proper cartographic attribution. However, most of the mentioned requirements do not require judgement by a skilled analyst and thus, are easy to automate. Furthermore, redundant features and attributes can also be easily identified and reported to the hydrographic data reviewer.

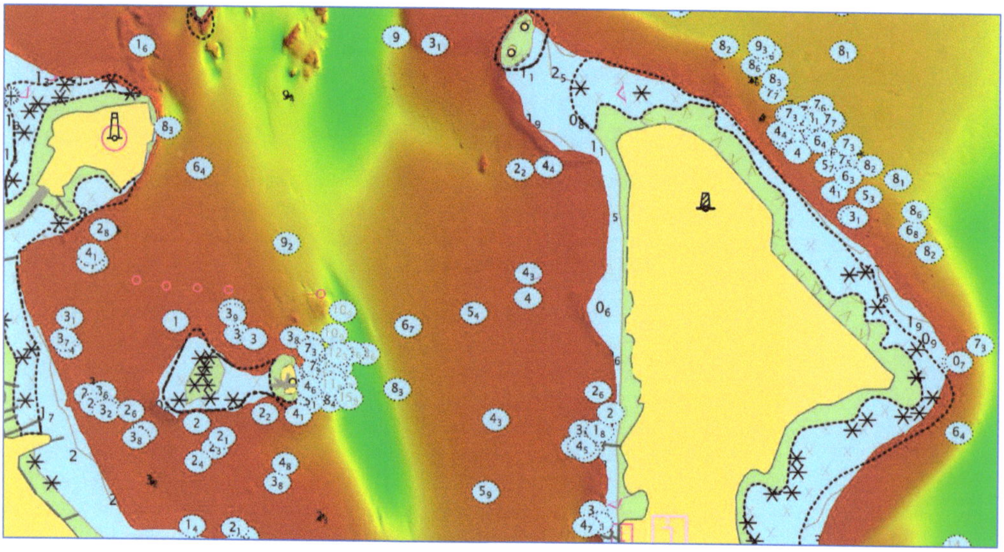

Figure 2. In nautical chart updates, the sheer number of features (represented by light blue circles, with the feature least depth sounding displayed inside) in nearshore areas is a task poorly befitting a manual review and is greatly aided by automation. Shown here is an Electronic Navigational Chart (ENC) US5NYCFJ, depicting part of the Western Long Island Sound, New York, NY, USA, with prospective chart features overlain atop gridded multibeam bathymetry (both from NOAA hydrographic survey H13384), which is colored by depth. All soundings are in meters; when shown, the sub-index represents decimeters.

Finally, significant features with an associated depth can be evaluated against the bathymetric grid to ensure that the grid and the feature attributes are consistent [13]. This latter task may appear simple, but the required amount of time quickly increases in nearshore areas saturated with features [17].

2.3. Survey Soundings and Chart Adequacy

High-density hydrographic surveys commonly consist of millions of survey soundings [1,7]. A bathymetric grid may be seen as a spatial filter for those hydrographic datasets to reduce the number of soundings based on reliable criteria. To preserve the safety of navigation, a common requirement is to assign the shoalest depth value among all the soundings within each grid cell [29]. However, gridding is just one of the possible methods used to identify a meaningful subset of the survey dataset to be used for cartographic processes [30–32].

During the hydrographic data review, it is often necessary to compare two different sets of depth values, e.g., a sounding selection. A common requirement is to compare a dense selection from the hydrographic survey under review with a sparser set of soundings and depth-attributed features derived from the chart. From such a comparison, shoals and dangers to navigation can be easily identified [15]. A similar procedure can be used to validate a set of newly proposed charted soundings against the original dense survey dataset. In both cases, the denser of the two sets may normally consist of tens of thousands of soundings, thus the manual execution of a similar task by the reviewer may end with several inconsistencies, some of them potentially associated with high safety-of-navigation risks [15]. As such, the development of automated procedures targeting the comparison of sets of depths has been critical for supporting the review process and specifically, to ensure that no critical shoal depths were missed.

3. Implementation and Results

In the past few years, the automated procedures outlined in the previous section have been implemented in two software applications, called QC Tools and CA Tools, developed in the HydrOffice framework [15,16]. HydrOffice (www.hydroffice.org, accessed on 8 August 2022) is an open-source collaborative project to develop a research software environment containing applications to strengthen all phases of the ping-to-public process in order to facilitate data acquisition, automate and enhance data processing, and improve hydrographic products [6].

QC Tools and CA Tools are currently implementing the survey specifications (i.e., the NOAA HSSD [13]) and other internal best practices of the NOAA Office of Coast Survey. Both tools are publicly available in Pydro—a free and open Python distribution—and as stand-alone applications (downloadable from the HydrOffice website: https://www.hydroffice.org/qctools/main, accessed on 8 August 2022; and https://www.hydroffice.org/catools/main, accessed on 8 August 2022) [33]. The stand-alone applications are currently distributed only for Microsoft Windows, although the underlying source code is cross-platform (e.g., Linux).

The algorithmic interpretation of the Office of Coast Survey's directives in both tools is regularly updated to reflect relevant changes introduced by the agency. The tools are also useful to train new personnel by helping them identify grid inconsistencies and feature issues, as well as in the interpretation of the survey specifications.

The code base of both software tools is similarly organized, consisting of a library, where the algorithms are implemented, and mechanisms to access such a library:

- Several scripts that can be used as a foundation to create new, custom algorithms.
- A command line interface useful to integrate some of the algorithms in the processing pipeline.
- An application with a graphical user interface (the *app*).

Both apps have a similar design to ease the user experience: they are arranged with a few main tabs and several sub-tabs. Specifically, the QC Tools app is organized into three main tabs. The first two are the *Survey Validation* tab and the *Chart Review* tab; these provide access to the QC tools themselves. The CA Tools app is organized into two main tabs, with the first one being the *Chart Adequacy* tab, providing access to the chart adequacy tools. Finally, for both apps, the last tab (the *Info* tab) includes support material, such as access to offline/online documentation and license information.

3.1. QC Tools

QC Tools provides automated procedures to:

- Detect candidate fliers and significant holidays in gridded bathymetry.
- Ensure that gridded bathymetry fulfills statistical requirements (e.g., sounding density and uncertainty).
- Check the validity of BAG files containing gridded bathymetry.
- Scan selected designated soundings to ensure their significance.
- Validate the attributes of significant features.
- Ensure consistency between grids and significant features.
- Extract seabed area characteristics for public distribution.
- Analyze the folder structure of a survey dataset for proper archival.

3.1.1. Grid Quality Controls

The *Detect Fliers* tool, also known as *Flier Finder*, aims to identify potential fliers in dense bathymetric grids. As previously mentioned, fliers can come in different types. As such, seven distinct algorithms have been developed over the past several years (see Table 1). Some of the algorithms require a search height as a parameter. When required by the algorithm, the search height may be used to tune the sensitivity to potential anomalies. For instance, the optimal search height on a relatively flat seafloor and shallow waters is usually smaller than for a dynamic area covered by a deep-water survey. Although the search height may be manually defined by the reviewer, the suggested solution is to have it automatically derived by an internal algorithm implementing a heuristic approach function of the median depth, depth variability, and grid roughness. The automated estimation of the search height helps standardize the hydrographic data review.

Table 1. Algorithms currently in use by the *Detect Fliers* tool.

Detect Fliers' Algorithm	Search Height Required
Laplacian Operator	Yes
Gaussian Curvature	No
Adjacent Cells	Yes
Edge Slivers	Yes
Isolated Nodes	No
Noisy Edges	No
Noisy Margins	No

The *Laplacian Operator* (Figure 3), the *Gaussian Curvature* (Figure 4), and the *Adjacent Cells* algorithms aim to detect shoal or deep spikes throughout the entirety of the bathymetric grid, whereas the *Edge Slivers* algorithm identifies potential fliers—mainly due to sparse data—on grid edges. The *Isolated Node* algorithm detects the presence of soundings detached from the main bathymetric grid that are often difficult to identify manually. Both the *Noisy Edges* (Figure 5) and *Noisy Margins* algorithms are tailored to identify fliers along noisy swath edges using the International Hydrographic Organization's S-44's Total Vertical Uncertainty (in place of the mentioned search height) [34]. The development of these latter algorithms was triggered by the fact that depth values associated with isolated nodes or on the grid edges are often unreliable when derived from the outmost beams of a bathymetric swath [35,36].

The *Detect Holidays* tool, also known as *Holiday Finder*, performs a grid search for holidays. The algorithm first identifies all the grid holidays, regardless of their size; then those holidays are tested against the survey specifications. Following the NOAA HSSD, the tool assess holidays based on the required survey coverage: either *Full Coverage* (Figure 6) or *Object Detection* (the latter having more restrictive criteria) [13]. The algorithm has been coded to calculate the holiday size (in number of nodes) based on the minimum allowable resolution and the grid resolution, but it is flexible for adjustments to different holiday descriptions.

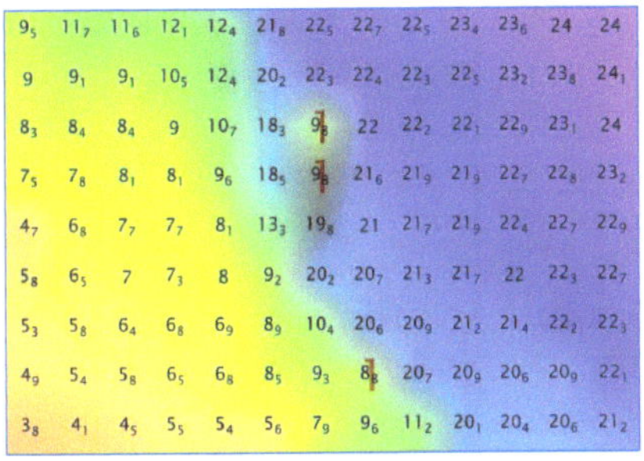

Figure 3. Example of potential fliers detected by the *Laplacian Operator* algorithm (marked with an orange 1). The black values are depth values in meters from the evaluated grid; when shown, the sub-index represents decimeters. The algorithm calculates the Laplacian operator as a measure of curvature by summing the depth gradients of the adjacent nodes. A cell is flagged as a potential flier when the resulting absolute value is greater than the search height.

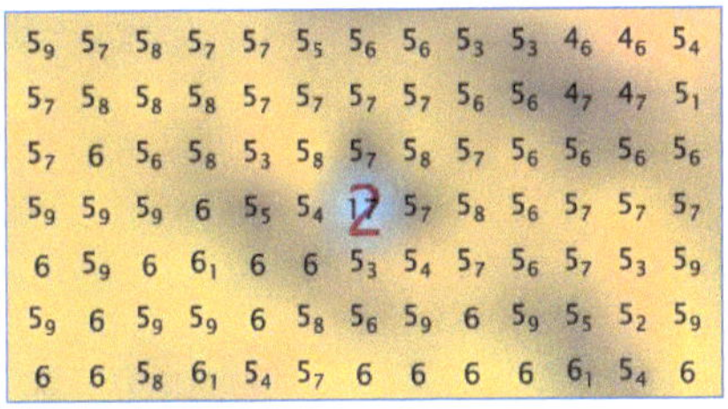

Figure 4. Example of a potential flier detected by the *Gaussian Curvature* algorithm (marked with an orange 2). The black values are depth values in meters from the evaluated grid; when shown, the sub-index represents decimeters. The algorithm bases the detection of potential fliers on the Gaussian curvature as a measure of the concavity at each node.

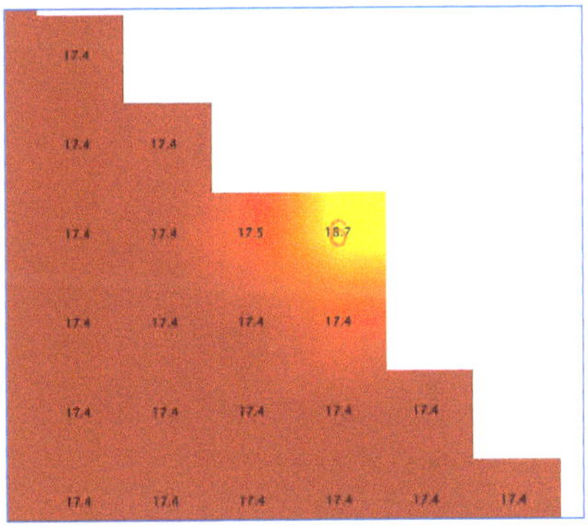

Figure 5. Example of a potential flier detected by the *Noisy Edges* algorithm (marked with an orange 6). The black values are depth values in meters from the evaluated grid; the sub-index represents decimeters. The algorithm crawls across empty cells to establish the edge nodes. Once an edge node is identified, the least depth and the maximum difference from its neighbors are calculated. The least depth is used to calculate to local Total Vertical Uncertainty, which is used for the flagging threshold [34].

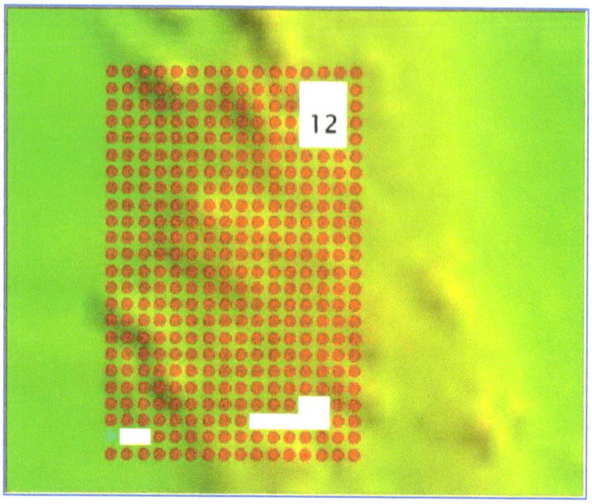

Figure 6. Outcomes from the *Detect Holidays* algorithm. The cells are marked with orange dots. The white areas are the grid gaps. Based on *Full Coverage* requirements, the gap of 12 grid nodes (black number) is marked as a holiday if it contains an instance of 3×3 unpopulated grid nodes [13]. The holes (white areas) with 7 nodes and 2 nodes do not fulfill such specifications.

The *Grid QA* tool performs statistical analysis on the bathymetric grid, looking at metrics such as data density (Figure 7), uncertainty (Figure 8), and, for variable-resolution grids, resolution requirements (Figure 9). Similar to the *Detect Holidays* tool, the current requirements are based on the NOAA HSSD [13], but can be adjusted to meet other specifications.

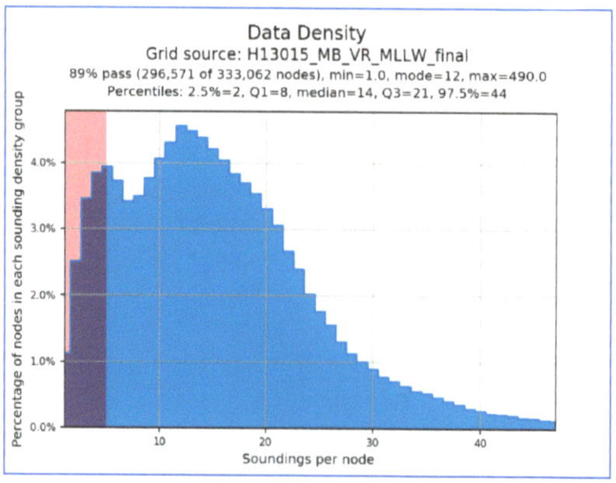

Figure 7. *Grid QA* output for data density. The histogram shows the percentage of total nodes that contain a specific sounding per node. To pass the density test, 95% of the nodes must have at least 5 soundings contributing to the population of that node [13]. The histogram bins with less than 5 soundings are in red. Therefore, in this example, this grid does not pass the density test; as noted in the title section of the figure, only 89% of the nodes pass this test.

The *BAG Checks* tool ensures compliance with the Open Navigation Surface Bathymetry Attributed Grid (BAG) format [18] for gridded bathymetry and, if selected, for additional NOAA-specific requirements. The algorithm checks the overall structure of the file, the metadata content, the elevation layer, the uncertainty layer, and the tracking list (an example of output is provided in Figure 10). It also performs a compatibility check with the popular GDAL software library and tools [37].

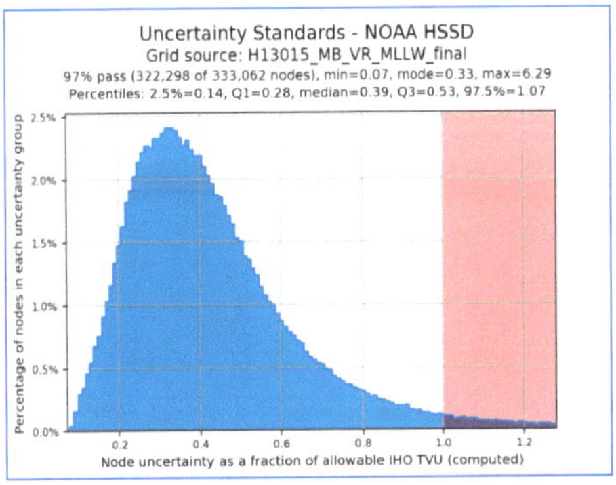

Figure 8. *Grid QA* output for uncertainty. The histogram illustrates the percentage of total nodes that contain a node uncertainty as a fraction of the International Hydrographic Organization's Total Vertical Uncertainty. As such, the histogram bins over 1.0 (in red) do not pass uncertainty requirements.

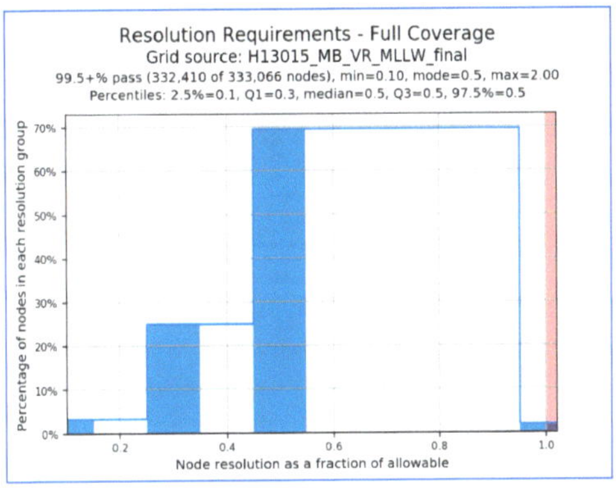

Figure 9. *Grid QA* output for resolution. Created only for variable-resolution surfaces, the histogram helps to identify the percentage of nodes that have a node resolution as a fraction of the allowable resolution at that depth. Anything over 1.0 (in red) does not pass the uncertainty requirements.

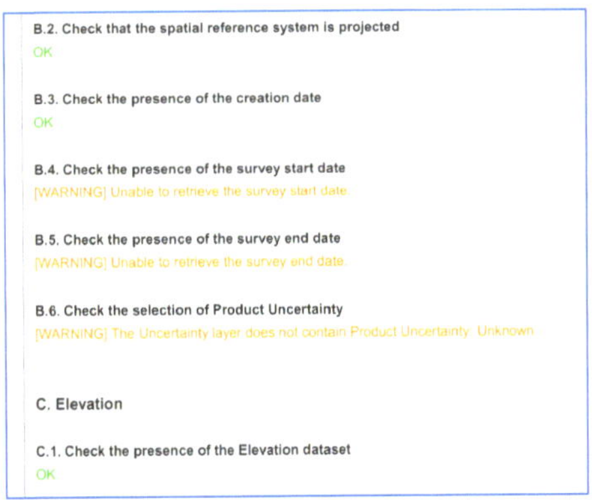
B.2. Check that the spatial reference system is projected
OK

B.3. Check the presence of the creation date
OK

B.4. Check the presence of the survey start date
[WARNING] Unable to retrieve the survey start date.

B.5. Check the presence of the survey end date
[WARNING] Unable to retrieve the survey end date.

B.6. Check the selection of Product Uncertainty
[WARNING] The Uncertainty layer does not contain Product Uncertainty: Unknown

C. Elevation

C.1. Check the presence of the Elevation dataset
OK

Figure 10. Extract from a PDF report generated by the *BAG Checks* tool. The report indicates which checks were performed and the results of the checks (passed checks in green, warnings in orange). At the end of the report, a summary indicates how many warnings and errors were identified for the surface.

The *Scan Designated* tool validates the soundings designated by the surveyor against the bathymetric grid to ensure their significance (according to NOAA HSSD specifications) [13]. Discrepancies are automatically highlighted for the reviewer (see Figure 11).

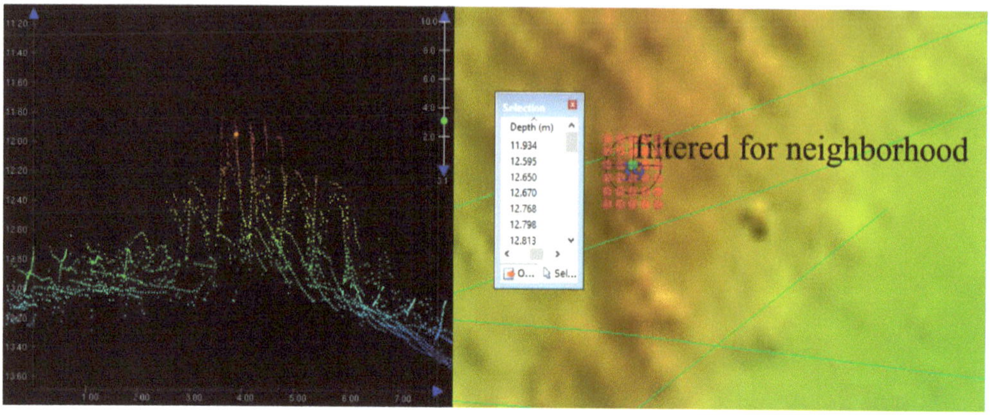

Figure 11. Example of Scan Designated output. The designated sounding appears less than 1 m off the seafloor when viewed in both sounding view (in the **left** pane) and grid data (in the **right** pane).

3.1.2. Significant Features Validation

The *Scan Features* tool checks the required S-57 attribution (e.g., [13]) for features that will be passed through the charting pipeline after the hydrographic data review (an example output report is shown in Figure 12). The tool provides several options to tailor the result to specific needs. For example, it is possible to switch between a *field profile* and an *office profile* based on the stage of the review pipeline at which the tool is executed. Other useful options are the version of the specification to be applied and additional checks, such as the image file naming convention, or the format of specific attributes (e.g., the date and the identification of the survey).

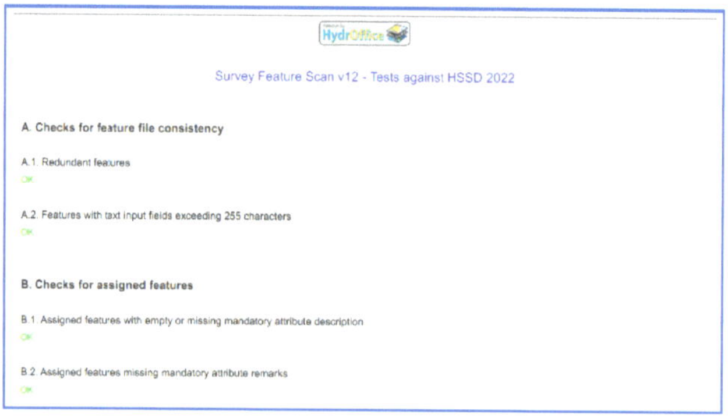

Figure 12. *Feature Scan* produces a PDF report that indicates which checks were performed and the results of the checks. At the end of the report, a summary indicates how many warnings and errors were identified, grouped by type.

The *Check VALSOU* tool evaluates all features against the corresponding grid nodes to ensure that the value of the sounding (VALSOU) and position matches what is present in the bathymetric grid. This tool not only ensures parity between feature depth and the grid, but it will also ensure that the depth entered is the most shoal depth among the nine grid nodes of the feature (see Figure 13).

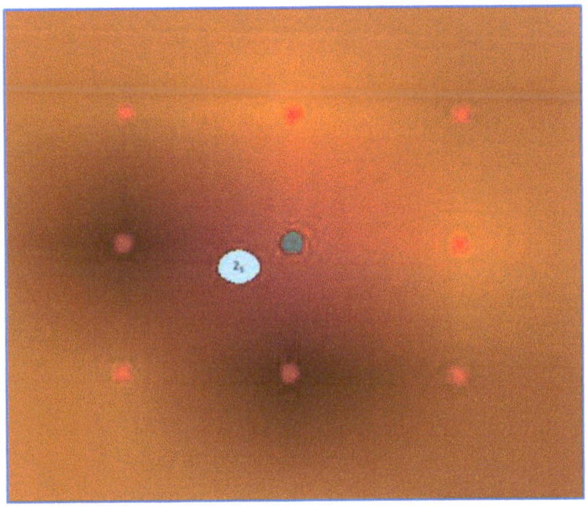

Figure 13. The *Check VALSOU* algorithm checks the grid node closest in position (cyan dot) to each significant feature and the eight grid nodes surrounding it (orange dots). The minimum depth value of one of these nodes must match the depth reported in the attribution of the significant feature.

3.2. CA Tools

CA Tools provides automated procedures to:

- Identify chart discrepancies for a bathymetric grid or a set of survey soundings.
- Select a significant set of soundings from a bathymetric grid.

The first step of the *Chart Adequacy* algorithm is to build a triangulated irregular network (TIN) from existing chart soundings and features; then it matches the dense set of survey soundings within the triangles of the TIN. At this point, the algorithm may apply two different testing methods: the *Shoalest Depth* method and the *Tilted Triangle* method. The *Shoalest Depth* testing method implements a longstanding Office of Coast Survey's best practice (called *Triangle Rule*) for the comparison of sounding sets (see Figure 14, pane A). In practice, any survey sounding shoaler than any of the three vertices of its containing triangle is marked as a potential problem. To overcome the inherent limitations of the Triangle Rule, the tilted-triangle test described in [6] (Figure 14, pane B) has been made available as the *Triangle Rule* testing method (see Figure 15). Due to the complexity of nautical charts, the algorithm also enforces additional sounding-in specific-feature tests [6]. The algorithm also computes the magnitude of the discrepancy against the chart and adds it as an S-57 attribute, allowing the identified soundings to be sorted. In this manner, the most significant discrepancies (and potential dangers to navigation) are identified immediately.

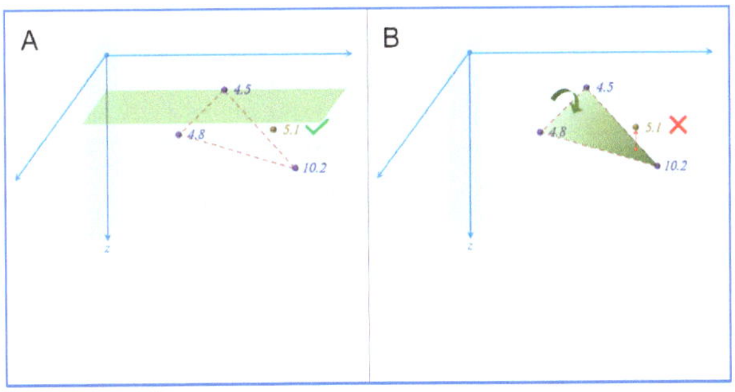

Figure 14. Example of the application of the *Shoalest Depth* testing method (i.e., the traditional *flat triangle test*) (**A**) and the *Tilted Triangle* testing method (**B**). The 5.1-m survey soundings (in dark yellow) are only flagged by the *Tilted Triangle* testing method when compared to the chart soundings (in purple).

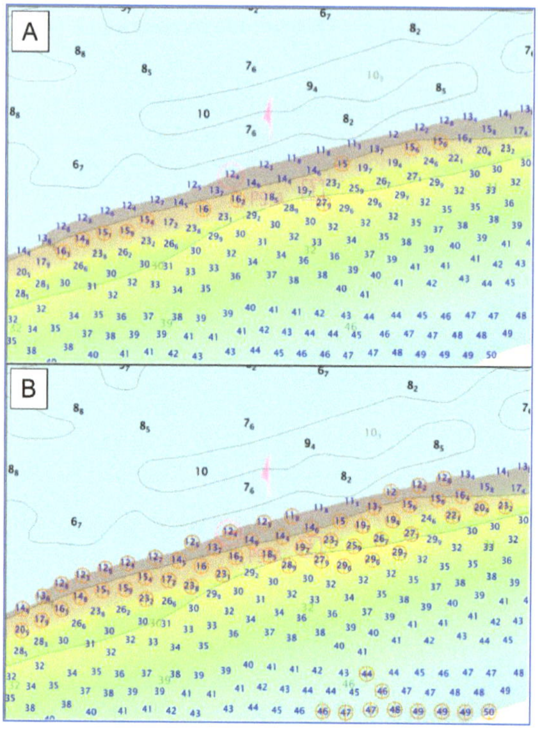

Figure 15. *Chart Adequacy* output using different testing methods. Chart soundings are shown in black and the survey soundings in blue. Both soundings are in meters; when shown, the sub-index represents decimeters. (**A**) shows the output from the *Shoalest Depth* method, only showing shoal soundings on the deep side of the contours. Thus, this method is useful in the identification of dangers to navigation. (**B**) shows the results using the *Tilted Triangle* method. There are more flagged soundings, in this case depicting the overall shoaling trend. Thus, this method is useful in change detection and assessing chart adequacy.

To summarize, the *Chart Adequacy* tool implements a method of sounding comparison that has two distinct applications: hydrographic survey review (as a quick identification of dangers to navigation) and chart review (as a method of validating a prospective chart sounding selection prior to its application).

The *Sounding Selection* tool creates a sounding selection from a bathymetric grid. Once created, the sounding selection can also be used to compare the survey data to the chart using the described *Chart Adequacy* tool. In fact, the initial motivation to create such a tool was to provide a mechanism to evaluate chart adequacy directly from a bathymetric grid. Two sounding selection algorithms are currently available: *Moving Window* and *Point Additive*. The *Moving Windows* algorithm is quite simple: the bathymetric grid is divided in square areas based on the size of a user-defined search radius (Figure 16, pane A); then the shallowest depth is selected within each area (Figure 16, pane B). The *Point Additive* algorithm iteratively selects the shallowest point in a bathymetric grid and then removes all cells within the radius of the selected sounding (Figure 17). The iteration continues until there are no remaining data points.

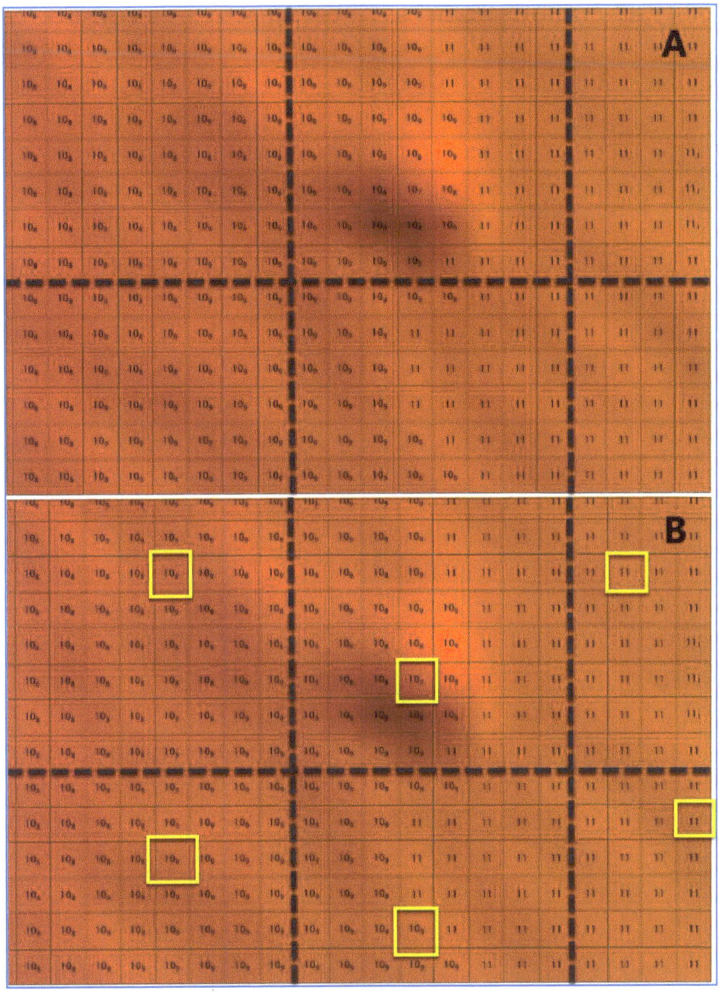

Figure 16. The *Moving Window* method used in the *Sounding Selection* tool. First, the area is divided into square window (**A**). The shallowest sounding is then chosen for each area (**B**). The black values are depth values, in meters, from the evaluated grid; when shown, the sub-index represents decimeters.

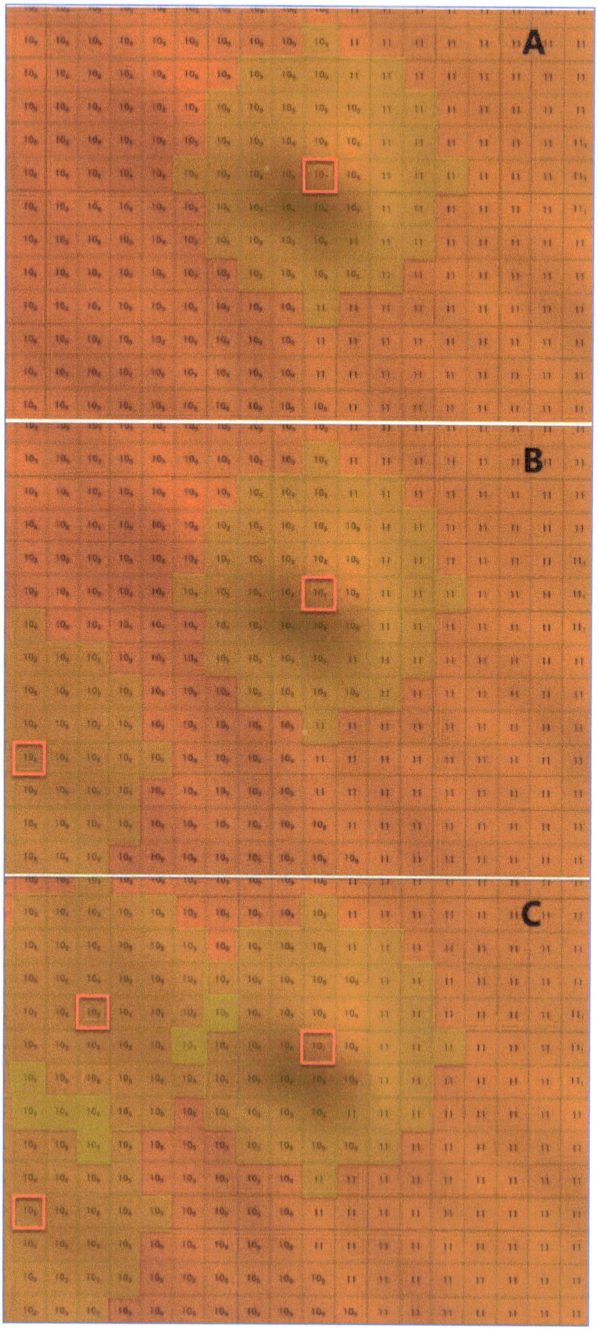

Figure 17. The *Point Additive* method used in the *Sounding Selection* tool. First, the shallowest sounding is selected, and the radius of soundings are removed (**A**). The next shallowest sounding is then chosen, radius removes neighboring soundings (**B**), and the process continues until all soundings are accounted for. The area of removed neighboring soundings can overlap, as in (**C**). The black values are depth values, in meters, from the evaluated grid; when shown, the sub-index represents decimeters.

4. Discussion

Applied to a large number of hydrographic surveys in recent years, the automated procedures in HydrOffice QC Tools and CA Tools have been shown to improve both the quality and timeliness of the review process [6,26]. An increased confidence in the final data produced was also observed, especially among personnel in training [6]. As such, the combined effect of applying these procedures is a novel holistic approach to hydrographic data review.

Both tools focus on several challenges present in the ping-to-public workflow, adopting a divide et impera (divide and conquer) approach and tackling the most time critical and error-prone steps [6]. By design, these tools are intended to be complementary to an existing hydrographic processing pipeline, providing valuable, and sometimes critical, supplementation of operator assessment with automated scanning over large datasets.

Although tailored to NOAA's processing and validation chain, the automated procedures are generically applicable to other hydrographic offices. The modular structure, inherited from the HydrOffice architecture, allows for the customization of the algorithms to different survey specifications. Furthermore, given that the code is neatly separated from the graphical user interface, the creation of stand-alone scripts is simple, both for local and cloud-based execution. For similar reasons, the code implementation of the specifications can be easily updated as the directives evolve.

These tools provide solutions for cases where software manufacturers are unable, or unwilling, to support the level of customization required by the hydrographic office. At the same time, these tools unambiguously provide algorithmic interpretation and evaluation of survey specifications. With a strong foundation of version-controlled algorithms, these tools represent a solid base for expanding automation in the future.

The feedback from the users within NOAA is positive, with the project receiving enthusiastic reviews from users, in terms of both frequency of use (Figure 18) and general evaluation (Figure 19) [6]. Furthermore, recently observed improvements in the Office of Coast Survey's data quality and timeliness has been partially attributed to the field implementation of these tools [3]. Given the similarities of review procedures among hydrographic offices, the described approach has generated interest in the ocean mapping community. This is mainly because the extent of the algorithmic interpretation of agency specifications represents the foundation for the adoption of automated workflows [16].

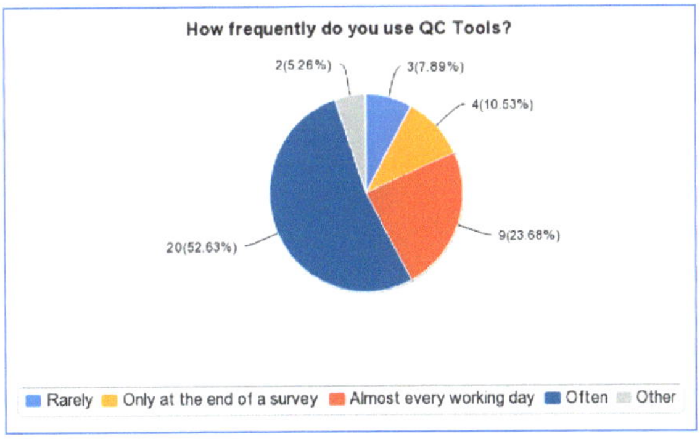

Figure 18. Customer satisfaction survey on QC Tools: frequency of use. Of the 39 survey respondents, more than 75% use QC Tools "often" or "almost every single working day" (more details are available in [6]).

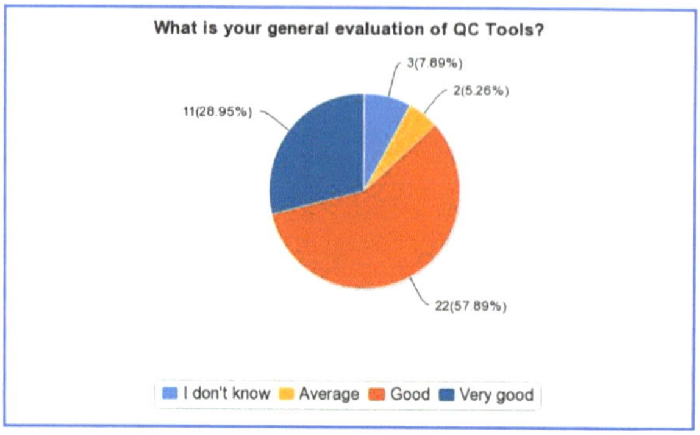

Figure 19. Customer satisfaction survey on QC Tools: general evaluation. A percentage larger than 86% of the survey respondents provide a general evaluation of the application as "good" or "very good" (more details in [6]).

A known limitation shared across the current implementations of both QC Tools and CA Tools is that visualizing their output requires an external GIS application that supports open hydrographic formats, such as BAG and S-57. Although most hydrographic software packages can read these formats, there are intrinsic limitations regarding how data reviewers can interact with the output. A possible solution to such an issue may be the creation of a plugin to interface the algorithm with an open GIS software, such as QGIS [38]. Such a solution will be explored as part of future development efforts.

Author Contributions: Conceptualization, methodology, and software, all authors; formal analysis, G.M.; validation, investigation, and resources, T.F., M.W., and J.W.; writing—original draft preparation, G.M.; writing—review and editing, T.F., M.W., and J.W. All authors have read and agreed to the published version of the manuscript.

Funding: This research was partially funded by the NOAA, grant numbers NA10NOS4000073, NA15NOS4000200, and NA20NOS4000196.

Data Availability Statement: The source code—with example scripts and data samples—is publicly available at: https://github.com/hydroffice (accessed on 8 August 2022). Future updates on the described initiative can be retrieved at: https://www.hydroffice.org (accessed on 8 August 2022).

Acknowledgments: Many keen hydrographic analysts and researchers from all around the world indirectly contributed to this work by providing feedback and endless sources of inspiration. In particular, the authors would like to thank the NOAA Coast Survey and the UNH CCOM/JHC for actively supporting their ideas for innovation. Finally, we also acknowledge the Coast Survey's Hydrographic Systems and Technology Branch for the continuous help in the integration within the Pydro software distribution.

Conflicts of Interest: The authors declare no conflict of interest.

References

1. Le Deunf, J.; Debese, N.; Schmitt, T.; Billot, R. A Review of Data Cleaning Approaches in a Hydrographic Framework with a Focus on Bathymetric Multibeam Echosounder Datasets. *Geosciences* **2020**, *10*, 254. [CrossRef]
2. Wlodarczyk-Sielicka, M.; Blaszczak-Bak, W. Processing of Bathymetric Data: The Fusion of New Reduction Methods for Spatial Big Data. *Sensors* **2020**, *20*, 6207. [CrossRef] [PubMed]
3. Evans, B. What are our Shared Challenges. In Proceedings of the NOAA Field Procedures Workshop, Virginia Beach, VA, USA, 24–26 January 2017.
4. Calder, B. Multi-algorithm swath consistency detection for multibeam echosounder data. *Int. Hydrogr. Rev.* **2007**, *8*. Available online: https://journals.lib.unb.ca/index.php/ihr/article/view/20778 (accessed on 8 August 2022).
5. Deunf, J.L.; Khannoussi, A.; Lecornu, L.; Meyer, P.; Puentes, J. Automatic Data Quality Assessment of Hydrographic Surveys Taking Into Account Experts' Preferences. In Proceedings of the OCEANS 2021: San Diego–Porto, Porto, Portugal, 20–23 September 2021; pp. 1–10. [CrossRef]

6. Masetti, G.; Faulkes, T.; Kastrisios, C. Hydrographic Survey Validation and Chart Adequacy Assessment Using Automated Solutions. In Proceedings of the US Hydro 2019, Biloxi, MS, USA, 19–21 March 2019. [CrossRef]
7. Hughes Clarke, J.E.; Mayer, L.A.; Wells, D.E. Shallow-water imaging multibeam sonars: A new tool for investigating seafloor processes in the coastal zone and on the continental shelf. *Mar. Geophys. Res.* **1996**, *18*, 607–629. [CrossRef]
8. Ladner, R.W.; Elmore, P.; Perkins, A.L.; Bourgeois, B.; Avera, W. Automated cleaning and uncertainty attribution of archival bathymetry based on a priori knowledge. *Mar. Geophys. Res.* **2017**, *38*, 291–301. [CrossRef]
9. Eeg, J. On the identification of spikes in soundings. *Int. Hydrogr. Rev.* **1995**, *72*, 33–41.
10. Debese, N.; Bisquay, H. Automatic detection of punctual errors in multibeam data using a robust estimator. *Int. Hydrogr. Rev.* **1999**, *76*, 49–63.
11. Hughes Clarke, J.E. The Impact of Acoustic Imaging Geometry on the Fidelity of Seabed Bathymetric Models. *Geosciences* **2018**, *8*, 109. [CrossRef]
12. Bottelier, P.; Briese, C.; Hennis, N.; Lindenbergh, R.; Pfeifer, N. Distinguishing features from outliers in automatic Kriging-based filtering of MBES data: A comparative study. In *Geostatistics for Environmental Applications*; Springer: Berlin/Heidelberg, Germany, 2005; pp. 403–414. [CrossRef]
13. NOAA. *Hydrographic Surveys Specifications and Deliverables*; National Oceanic and Atmospheric Administration, National Ocean Service: Silver Spring, MD, USA, 2022.
14. Jakobsson, M.; Calder, B.; Mayer, L. On the effect of random errors in gridded bathymetric compilations. *J. Geophys. Res. Solid Earth* **2002**, *107*, ETG 14-1–ETG 14-11. [CrossRef]
15. Masetti, G.; Faulkes, T.; Kastrisios, C. Automated Identification of Discrepancies between Nautical Charts and Survey Soundings. *ISPRS Int. J. Geo-Inf.* **2018**, *7*, 392. [CrossRef]
16. Wilson, M.; Masetti, G.; Calder, B.R. Automated Tools to Improve the Ping-to-Chart Workflow. *Int. Hydrogr. Rev.* **2017**, *17*, 21–30.
17. IHO. *S-57: Transfer Standard for Digital Hydrographic Data*; International Hydrographic Organization: Monte Carlo, Monaco, 2000.
18. Calder, B.; Byrne, S.; Lamey, B.; Brennan, R.T.; Case, J.D.; Fabre, D.; Gallagher, B.; Ladner, R.W.; Moggert, F.; Paton, M. The open navigation surface project. *Int. Hydrogr. Rev.* **2005**, *6*, 9–18.
19. Quick, L.; Foster, B.; Hart, K. CARIS: Managing bathymetric metadata from "Ping" to Chart. In Proceedings of the OCEANS 2009, Biloxi, MS, USA, 26–29 October 2009; pp. 1–9. [CrossRef]
20. Younkin, E. Kluster: Distributed Multibeam Processing System in the Pangeo Ecosystem. In Proceedings of the OCEANS 2021: San Diego–Porto, Porto, Portugal, 20–23 September 2021; pp. 1–6. [CrossRef]
21. van Rossum, G. *The Python Language Reference: Release 3.6.4*; 12th Media Services: Suwanee, GA, USA, 2018; p. 168.
22. Calder, B.; Mayer, L. Robust Automatic Multi-beam Bathymetric Processing. In Proceedings of the US Hydro 2001, Norfolk, VA, USA, 22–24 May 2001.
23. Hou, T.; Huff, L.C.; Mayer, L.A. Automatic detection of outliers in multibeam echo sounding data. In Proceedings of the US Hydro 2001, Norfolk, VA, USA, 22–24 May 2001.
24. Mayer, L.A. Frontiers in Seafloor Mapping and Visualization. *Mar. Geophys. Res.* **2006**, *27*, 7–17. [CrossRef]
25. Mayer, L.A.; Paton, M.; Gee, L.; Gardner, S.V.; Ware, C. Interactive 3-D visualization: A tool for seafloor navigation, exploration and engineering. In Proceedings of the OCEANS 2000 MTS/IEEE Conference and Exhibition, Conference Proceedings (Cat. No.00CH37158). Providence, RI, USA, 11–14 September 2000; Volume 912, pp. 913–919. [CrossRef]
26. Gonsalves, M. Survey Wellness. In Proceedings of the NOAA Coast Survey Field Procedures Workshop, Virginia Beach, VA, USA, 27–20 January 2015.
27. Briggs, K.B.; Lyons, A.P.; Pouliquen, E.; Mayer, L.A.; Richardson, M.D. *Seafloor Roughness, Sediment Grain Size, and Temporal Stability*; Naval Research Lab: Washington, DC, USA, 2005.
28. Hare, R.; Eakins, B.; Amante, C. Modelling bathymetric uncertainty. *Int. Hydrogr. Rev.* **2011**, *9*, 31–42.
29. Armstrong, A.A.; Huff, L.C.; Glang, G.F. New technology for shallow water hydrographic surveys. *Int. Hydrogr. Rev.* **1998**, *2*, 27–41.
30. Dyer, N.; Kastrisios, C.; De Floriani, L. Label-based generalization of bathymetry data for hydrographic sounding selection. *Cartogr. Geogr. Inf. Sci.* **2022**, *49*, 338–353. [CrossRef]
31. Zoraster, S.; Bayer, S. Automated cartographic sounding selection. *Int. Hydrogr. Rev.* **1992**, *1*, 103–116.
32. Sui, H.; Zhu, X.; Zhang, A. A System for Fast Cartographic Sounding Selection. *Mar. Geod.* **2005**, *28*, 159–165. [CrossRef]
33. Riley, J.; Gallagher, B.; Noll, G. Hydrographic Data Integration with PYDRO. In Proceedings of the 2nd International Conference on High Resolution Survey in Shallow Water, Portsmouth, NH, USA, 24–27 September 2001.
34. IHO. *S-44: Standards for Hydrographic Surveys*; International Hydrographic Organization: Monte Carlo, Monaco, 2020.
35. Hughes Clarke, J.E. Multibeam Echosounders. In *Submarine Geomorphology*; Micallef, A., Krastel, S., Savini, A., Eds.; Springer International Publishing: Cham, Switzerland, 2018; pp. 25–41.
36. Lurton, X.; Augustin, J.-M. A measurement quality factor for swath bathymetry sounders. *IEEE J. Ocean. Eng.* **2010**, *35*, 852–862. [CrossRef]
37. Warmerdam, F. The Geospatial Data Abstraction Library. In *Open Source Approaches in Spatial Data Handling*; Hall, G.B., Leahy, M.G., Eds.; Springer: Berlin/Heidelberg, Germany, 2008; pp. 87–104.
38. QGIS.org. QGIS Geographic Information System. Available online: http://www.qgis.org/ (accessed on 14 August 2022).

geomatics

Article

Automating the Management of 300 Years of Ocean Mapping Effort in Order to Improve the Production of Nautical Cartography and Bathymetric Products: Shom's Téthys Workflow †

Julian Le Deunf [1,2,*], Thierry Schmitt [2], Yann Keramoal [2], Ronan Jarno [2] and Morvan Fally [2]

1 IMT Atlantique, Lab-STICC, UMR CNRS 6285, F-29238 Brest, France
2 Service Hydrographique et Océanographique de la Marine (Shom), 13 rue du Châtellier, 29200 Brest, France
* Correspondence: julian.le.deunf@shom.fr
† This paper is extended from conference paper "Téthys: automating a data workflow compiling over 300 years of bathymetric information", San Diego, CA, USA, 20–23 September 2021.

Abstract: With more than 300 years of existence, Shom is the oldest active hydrographic service in the world. Compiling and deconflicting this much history automatically is a real challenge. This article will present the types of data Shom has to manipulate and the different steps of the workflow that allows Shom to compile over 300 years of bathymetric knowledge. The Téthys project for Shom will be presented in detail. The implementation of this type of process is a scientific, algorithmic, and infrastructure challenge.

Keywords: bathymetric data; quality analysis; data fusion and management; nautical cartography

Citation: Le Deunf, J.; Schmitt, T.; Keramoal, Y.; Jarno, R.; Fally, M. Automating the Management of 300 Years of Ocean Mapping Effort in Order to Improve the Production of Nautical Cartography and Bathymetric Products: Shom's Téthys Workflow. *Geomatics* **2023**, *3*, 239–249. https://doi.org/10.3390/geomatics3010013

Academic Editor: Gabriel Navarro

Received: 11 January 2023
Revised: 12 February 2023
Accepted: 20 February 2023
Published: 22 February 2023

1. Introduction

Since 1720, Shom, the French Hydrographic Service, has collected information describing the physical marine environment, including bathymetric (depth) measurements. These 300 years of data holdings originate from different types of sensors: either through mechanical means (lead-line from early 1800 to the 1920s) [1], or more commonly from acoustic sounding (single-beam since the 1820s, then using multi-beam sounders since the 1980s) [2] or even from optical sounding (lidar since 2005) [3]. The data acquired therefore has different characteristics and qualities. These are at the basis of nautical products, including nautical charts, ensuring the safety of navigation for the mariners, and compliance with Regulation 9 Chapter V of the SOLAS convention; see [4].

All these accumulated data are integrated into a single dedicated database, Shom's bathymetric database (SBDB), see Figure 1, managed as a pile of overlapping and/or intersecting surveys. As of 2022, the SBDB holds over 11,400 surveys. Currently, each cartographic operator that generates nautical charts or digital terrain models must go through a laborious process of selection of bathymetric information.

The Téthys project is an in-house project aimed at constituting Shom's bathymetric reference surface for which source data have been selected in order to generate the most accurate and up-to-date surface, satisfying the criteria related to the safety of navigation.

In this paper, we will first provide elements related to the purpose of bathymetric data fusion along with the current methodology for the production of nautical information. With this context being laid down, we will then explain the details of the Téthys project: notably the data model (especially the geographical extent of the data) and the semi-automated processing chain. Finally, through examples, we will present the current production status results, before proposing conclusive perspectives.

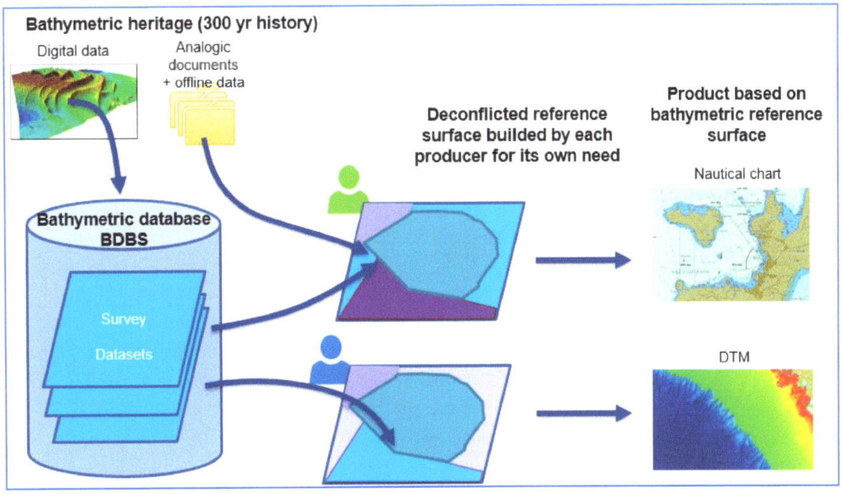

Figure 1. General operation before the Téthys project, where each operator recovers all the data before processing it according to his needs.

2. Bathymetric Data Fusion State of the Art

Bathymetric data fusion is a vast research topic, which has been largely discussed in the literature [5–8]. Globally speaking, the process deals with the aggregation of depth soundings for dedicated products. Depth soundings can originate from diverse sensors (acoustical: single beam or multibeam; estimated from active or passive remote sensing technologies: Lidar, satellite-derived bathymetry, Radar Altimetry) and prior information originating from historical digitized charted products or ENCs. These sources of data often combine multiple levels of vertical and horizontal accuracies; they integrate various levels of data density (from sparse coverage to tenths to a hundred of soundings per m^2) from different ages and different processing levels (from raw soundings to fully corrected and precisely vertically reference datasets associated with a characterization of the associated level of confidence).

Due to the scarcity and the difficulty of collecting bathymetric data, the users of bathymetric products generated through a process of data fusion are often aware that data with mixed levels of quality is the best that can be achieved. The scale and coverage of the generated product are often the key factors associated with the level of refinement and effort to be put into the compilation process [9]. At a small scale (i.e., roughly speaking above 1/10.000.000), common current practice is to aggregate all the sources and use deterministic interpolators (quite often spline-in-tension). This is exemplified by a worldwide digital terrain model such as SRTM15+ [10] or GEBCO [11]. Strong limitations arise, notably when the level of details and accuracy sought is higher (larger scale, higher resolution) in particular in dynamic areas (e.g., sandwaves).

For this kind of context, a selection process, focusing principally on the most recent data sources, is needed to prevent an unrealistic representation of the seabed. Regional efforts such as EMODnet Bathymetry incorporate this step in their methodology [12]. In shallow water areas, where vertical precision is essential, attention must be paid to the quality of the dataset with respect to vertical accuracy, with particular attention being paid to the tidal referencing methodology [13]. Some national data compilation efforts well illustrate this further effort in the data selection and management process [9,13,14].

Moreover, when it comes to bathymetric product usages where the security of navigation is at stake, all the previous steps of the selection processes are taken with extra caution, as much as the representation of the various confidence levels associated with the source data is included in the process of compilation. In this sense, the CATZOC (Category of the Zone of Confidence) is an accompanying layer of the navigational chart [15] describing the data uncertainty [16], which has as much importance as the depth information layer, and which is supported by dedicated and well-managed metadata associated to the source datasets (see Section 4).

A major keystone in the reasoning for generating bathymetric products from hydrographic data is the concept of "Navigation Surface," also known as reference surface, first introduced by Smith [17]. The navigation surface is a bathymetric surface product made up of a collection of sources assuring that items critical for navigation are preserved. One of the benefits of this concept is to locally ensure the selection of the most relevant data sources compatible with the most stringent use (here considered to be the safety of navigation).

The process of ensuring the suitable selection of the local knowledge is also known as "deconfliction," which is a decision-making process whereby a selection of the sources of data is made from the best representation of the physical coverage (2D polygon area) and associated information such as, for example, the age and reliability of the source. This process is either undertaken (see Figure 2) through:

- a decision of "remove/restore," where one of two (or many) overlapping sources uptakes the others.
- a decision of "supplement," where the datasets under concern are merged without giving priority (or weight) to one dataset over the other.

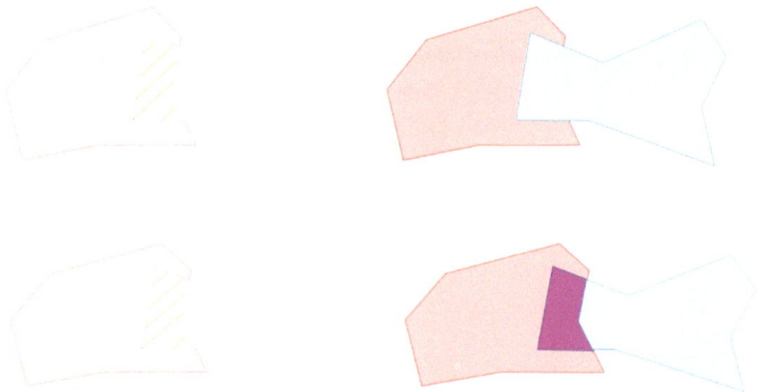

Figure 2. Deconfliction process representation. **Top**: a remove/restore decision where the blue survey uptakes the red one; **Bottom**: a supplement decision where both surveys are merged.

By creating such a navigation surface at the geographical scale of the survey (highest scale), the objective is to undertake the decision-making selection process once for all sthe cales. In doing so, gains are already measured in generating bathymetric products (nautical charts and digital terrain models) by capitalizing on the selection efforts while strengthening the management and valorization of the source information. Currently, other similar national initiatives are underway, such as BlueTopo from the NOAA [18].

3. Qualitative Description of Data and Metadata

The information used in the production of Shom's reference surface is composed of bathymetric data from hydrographic surveys (sounding point clouds in the form of x,y,z triplets or bathymetric raster surfaces) to which are associated spatial metadata representing the extent of the survey (minimal enclosing surface, later defined as MES and SME in French) and a series of attribute metadata.

The attribute metadata are associated with internationally recognized metadata from the IHO S-57 standard [19,20], such as CATZOC (Category Zone of Confidence), POSACC (Position accuracy), SOUACC (Sounding accuracy); see Figure 3 as well as internally defined attributes. From the latter, we can cite examples such as the Codval attribute, which indicates the validity/invalidity of a bathymetric survey, or the Captur attribute, which indicates the type of bathymetric sensors used at sea for its acquisition. Within the framework of the Téthys, the attribute metadata are managed in a conventional eXtended Markup Language (XML) format with a key-value formalism.


```xml
<?xml version="1.0" encoding="UTF-8"?>
<BDB_Simple_Attributes>
  <Attribute name="CATZOC">
    <Value>zone of confidence B</Value>
  </Attribute>
  <Attribute name="NINFOM">
    <Value>20210611 : STM/BATHY/GDBS Reprise de la SME. Création d'un fichier
&#13;
Jun 11, 2021 8:22:55 AM&#13;
Bathymetry replaced by blecorre</Value>
  </Attribute>
  <Attribute name="OBJNAM">
    <Value>S201500100-1</Value>
  </Attribute>
  <Attribute name="POSACC">
    <Value>2.1</Value>
  </Attribute>
  <Attribute name="RECDAT">
    <Value>20170216</Value>
  </Attribute>
  <Attribute name="SOUACC">
    <Value>0.43</Value>
  </Attribute>
```


Figure 3. Classic metadata file with S-57 [19] attributes.

The spatial metadata of a bathymetric survey is meant to represent the area of the seabed that has been recognized following the acquisition/processing stage. Fidelity to the actual coverage of the dataset is key as this polygon will be used as part of the deconfliction process. In order to be the most representative, three conditions on the relationship between the minimal enclosing surface polygon and its associated soundings have been defined and must be verified:

- Unicity: Each sounding of the dataset is included in a single MES.
- Density: Soundings that have a distance with the nearest neighbors less than 5 times (defined from the hydrographic expertise, Case 1 of Figure 4) the intrinsic resolution of the acquisition sensor is gathered in the same MES. Otherwise, a new polygon is created (Case 2 of Figure 4). Eventually, the sounding is considered as an isolated sounding if it is impossible to aggregate it with its neighbors (Case 3 of Figure 4).
- Representativeness: The contour (internal and external) of the generated multi-polygons is buffered with a distance depending on the characteristics of the survey: horizontal uncertainty and intrinsic resolution. This is a sensitive point to avoid removing a shoal at the border of the survey with a too-loose MES when the area has not been strictly covered.

Figure 4 schematizes all these criteria. Note that following this process, a single survey is represented by a single or a multi-polygon also including holes in their geometry.

Prior to 2018, the MES was constructed manually by operators at Shom, as no known algorithm provided satisfaction (representativity and computing performance). Moreover, this tedious work was also subject to operators' biases, which strongly motivated the development of an automated MES envelope generation.

In order to determine the most accurate spatial coverage for a bathymetric survey, we first studied the α-shape algorithm, which is a classical computational geometry method [21] that is a refinement of the convex hull method [22]; both are available in numerous GIS solutions (e.g., QGIS) based on a Delaunay Triangulation followed by an analysis of the length of the triangle edges and suppression of the triangle edges, the lengths of which are above the defined α length. The algorithm has a complexity of $O(n \log n)$, with n representing the number of points. The α-shape algorithm is used to obtain the line segments composing the perimeter of a set of points in the plane, thus allowing the building of the strictest spatial boundary from these segments composing the boundary of the input point cloud. The key parameter of this algorithm is the α value, which defines whether a segment will be considered a right-of-way boundary or a core segment. Nevertheless, this method has strong limitations, with changing density of the point cloud, considering the static definition of the α parameter.

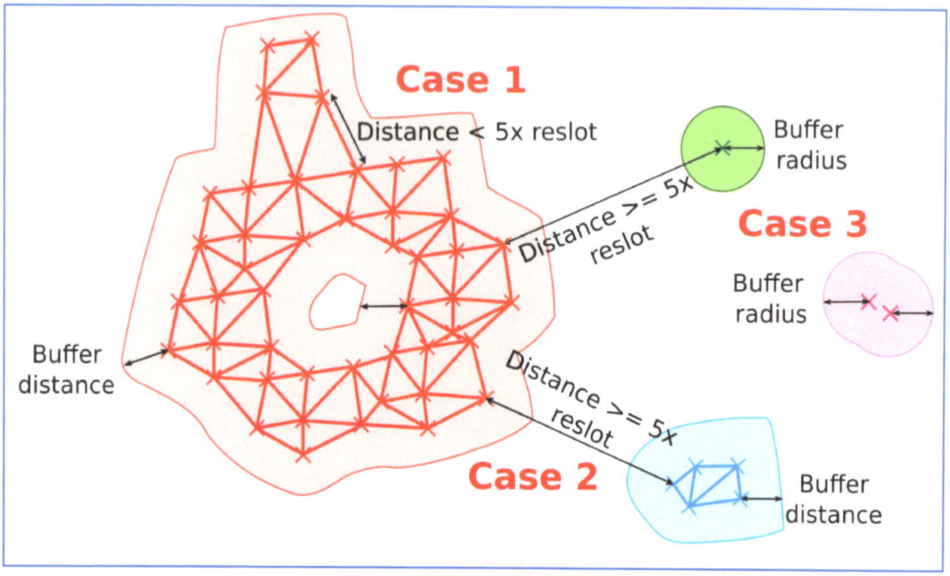

Figure 4. Shom MES definition.

In order to alleviate the limitation introduced by the α-shape algorithm and to optimize the searching phase of the points located at the border of the survey, we introduced the QuadSME algorithm [23]. Based on the bathymetric point cloud and its horizontal uncertainty (the POSACC), the algorithm had five steps:

- The first step of this methodology was the import of the data in the form of a point cloud including triples (x, y, z) and the value of the associated POSACC.
- The second step of the methodology consisted of a geospatial indexation based on a first quadtree segmentation [24], with the number of points per quadrant set to 5 million points as the stopping criterion.
- The third step of the methodology consisted of a second quadtree indexing. The space was divided to keep only quadrants validating either a density criterion or a maximum number of soundings (arbitrarily defined at 1000). The density criterion corresponded to the number of points in each sub-quadrant constituting a main quadrant. If the density was identical (judged by a threshold) for each child quadrant, then the parent quadrant was considered as homogeneous.
- The fourth step of the methodology consisted of the generation of polygons containing the soundings. First, a characteristic resolution of the point cloud included in the sub-quadrant was calculated to adapt to the potential differences in density of the input point cloud. Then, partitioning and detection of isolated points were performed. The objective was to build specific envelopes for the isolated points and build clusters of points with the same density before creating the polygons. Finally, a Delaunay triangulation was performed on the different clusters and the associated polygon was extracted.
- The fifth and last step of the methodology consisted of dissolving the polygons generated during the previous steps to form the final MES. The geometries were merged via a process of dilation/erosion (creation of a buffer) of the geometries to

remove construction holes, see Figure 5 which represents a generated MES and the associated quadtree decomposition.

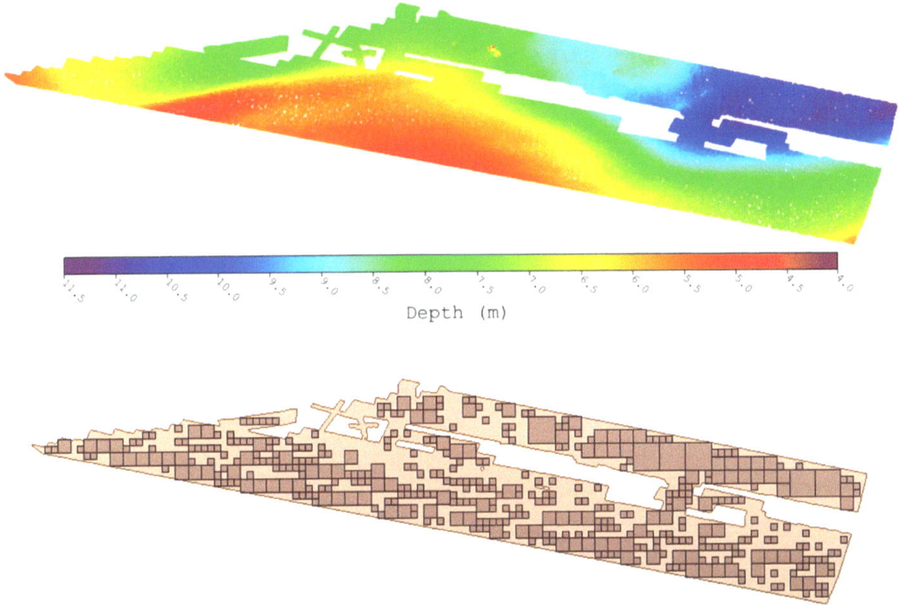

Figure 5. Top: a bathymetric survey (point cloud—color according to depth value); **Bottom**: the MES generated by the QuadSME algorithm with the preserved holes. Note also the corresponding quadtree decomposition.

In order to compare the representativity of the geometries both generated by the α-shape and QuadSME methods, the Haussdorf-Pompeiu [25] distance metric was selected. The Hausdorff-Pompeiu distance is a topological tool that measures the distance between two subsets of a metric space. It was therefore very suitable for comparing the maximum distance between two spatial areas, which allowed the dissimilarity of the two shapes to be measured. From Table 1, which shows the results associated with five surveys differing in size and geographical coverage, it can be observed that the Hausdorff-Pompeiu distance for the QuadSME method was always smaller and therefore more faithful to the reference α-shape method. Also, a fact to be noticed is that the distance value remained in the same order of magnitude for lots with few soundings (first two examples of Table 1). Moreover, for larger size datasets (last three examples of Table 1), the QuadSME method provided a Hausdorff-Pompeiu distance better within one order of magnitude.

Table 1. Computation of the Hausdorff-Pompeiu distance (in meters) for five bathymetric surveys.

Survey Name	Soundings Number	α -Shape Distance	QuadSME Distance
S202099900-001	2743	165.6	141.9
S201207000-5	25,718	86.7	76.5
E201804100-002	128,939	16.6	0.9
S202102500-001	1,070,131	182.9	25.1
S200701200-1	10,829,541	1340.3	118.3

The computation time associated with each method was also compared, using the same computing facility (Intel Xeon 6248 2.50 GHz, 32 Gb RAM). The QuadSME method, see Figure 6, showed better computation times than the QuadSME algorithm compared to the α-shape algorithm, especially when the number of points was greater than one million. For a number of soundings of the order of magnitude of 10 million, the QuadSME method was 40 times faster than the α-shape algorithm, most likely because of the quadtree partitioning (Steps 2 and 3). Processing time was further improved by multiprocessing the QuadSME method, with operations from Steps 2 to 4 performed independently and in parallel on each of the quadrants.

Figure 6. Computation time for α-shape and QuadSME algorithms.

On the other hand, the computation time of the QuadSME method was very dependent on the homogeneity of the distribution soundings. Thus, when the density criterion was quickly reached then the quadtree process stopped. Conversely, when the sounding distribution was not homogeneous in the sub-quadrants or when the data contained many holes, then the computation time was longer because it was necessary to go to the end of the quadtree decomposition.

4. Téthys Workflow

Following a detailed and accurate representation of the source dataset, as described in the previous section, the deconfliction process was wisely undertaken, leading to the generation of the bathymetric surface reference. The overall workflow, see Figure 7, was carried out according to the following processes:

- From the different original surveys, verification of all data and metadata content was performed, benefiting the SBDB consistency directly.
- The conflicts between the superimposed datasets were resolved according to the qualitative elements carried by the metadata (hydrographic qualification, ages, etc.).
- The compilation (combination or cutting/replacement) of the data was undertaken following the priorities previously defined between the datasets in Step 2.

Considering the vastness of the French exclusive economic zone (EEZ), this workflow was operated on 1° by 1° geographic tiles. More than 300 expert rules validated by Shom hydrographers and cartographers were implemented in this process.

The Téthys project offers each operator data where their interactions are validated and verified by a set of attribute rules and priority constraints related to each other. The resulting surface is directly exploitable, without any particular expertise, and is reproducible.

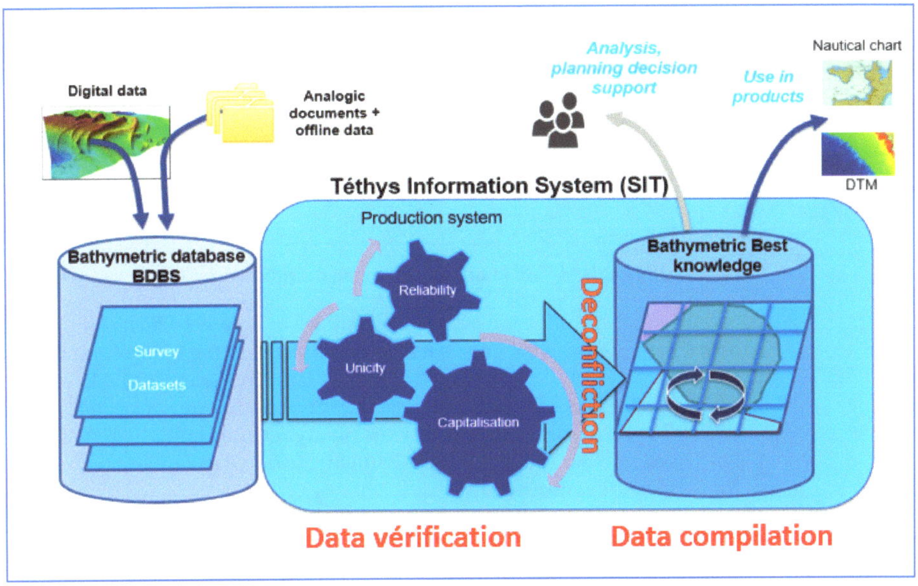

Figure 7. General workflow of Téthys Project, where the Téthys base is the reference bathymetric bottom.

Automation of this workflow can be implemented based on several technologies that best handle open and proprietary geospatial data formats, along with efficient manipulation of large volumes of data. Fundamental to this implementation are the use of:

- Extract Transform Load (ETL) software handling spatial information: The dedicated FME software [26] supports geospatial data extraction (Extract) from homogeneous or heterogeneous sources, followed by the processing stage (Transform) of the data into a proper storage format/structure; and, finally, the data is loaded into a dedicated target database. In addition to data transformation tools, spatial ETL solutions also contain various geoprocessing algorithms to process and analyze spatial and non-spatial data (e.g., geometry validation and repair, topology check, or creating and merging attributes, etc.). The software allows this tool to have several advantages for the needs of Téthys. Figure 8 illustrates the no-code graphical FME interface, based on multiple data-driven interactors. Such a workflow processing environment facilitates development and subsequent maintenance.
- Direct geo-processing in a dedicated working database via SQL scripts: The choice was made to use the combination of a PostgreSQL/POSTGIS database, overlaid with the pgPointCloud extension [27]. This environment benefits from the adapted geospatial point cloud indexing capabilities commonly used for the management of large LIDAR point clouds, which have similar characteristics to bathymetric soundings. Note that direct interaction with the pgPointCloud data structure is managed through the PDAL library [28].
- Dedicated APIs to allow for the manipulation of proprietary format. The current SBDB is currently managed under the proprietary software Teledyne CARIS Bathymetric DataBase, and Python bindings built upon a dedicated API provided by the software manufacturer [29] allow for the transformation into open and interoperable formats.

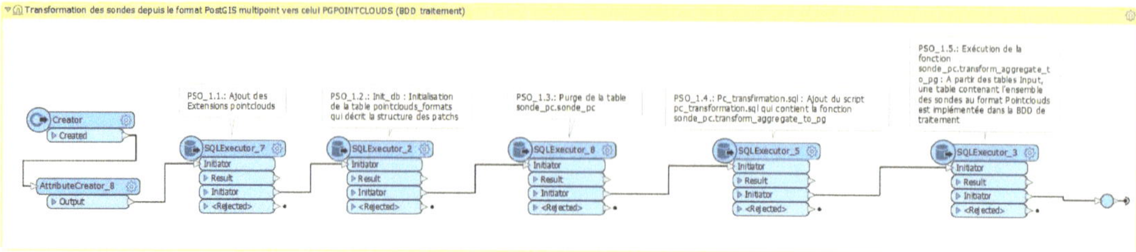

Figure 8. Example of FME interface performing the processing to generate the surface reference.

The first deconfliction performed is shown in Figure 9 which distinguished the stages before and after this process; each color represented the MSE of a survey. On this first tile (called 145_81, a name inspired by the Marsden square [30]), 115 surveys were used as input data and 441,418,088 associated soundings were processed. At the end of the processing chain, only 96 surveys were finally retained and 310,970,981 soundings were integrated into the Téthys. Of these bathymetric data, over 6000 soundings were digitized from old nautical charts.

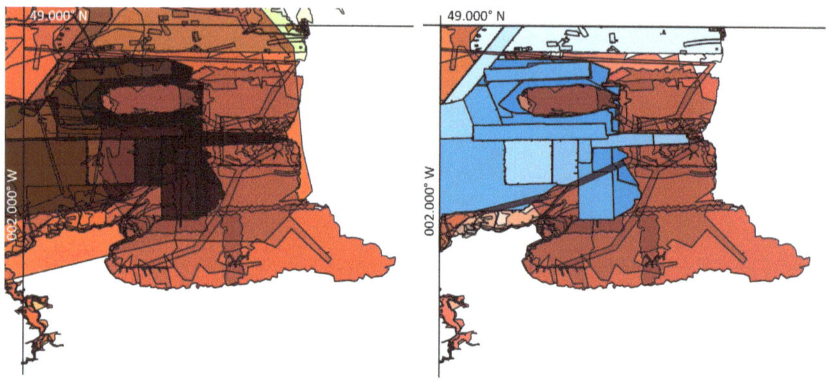

Figure 9. First tile of the Téthys project: **on the left**, surveys studied for deconfliction; **on the right**, surveys cut and kept after the deconfliction process (we distinguish easily the remove/restore or supplement decision especially in the blue part on the right side).

The area covered extends from the port of Saint-Malo in the west, to the bay of Mont Saint-Michel in the east, and from the south of the Rance to the town of Coutance in the north. The result of this deconfliction process raised 44,167 conflicts between intersecting data sources. Quality control of the tiles was performed by comparing, among other things, previously generated navigation products, such as the official Electronic Navigational Chart (ENC). This first work has recently benefited cartographers who published the nautical chart covering the Chausey Islands and the production of the topo-bathymetric DTM, which covers the approaches to Saint-Malo; see [30].

Shom agents have access to these bathymetric data via an internal geographic web portal. Selected layers can be queried, filtered, and downloaded in well-known GIS formats (ASCII, shapefile, GeoPackage). Bathymetric data are extracted by defining a bounding box of the area of interest. Users can also load web services (WMS, WFS, WCS) or GIS vector data into the portal.

5. Discussion and Perspectives

The current concept underlying the use of the Téthys is oriented towards cartographic use applied for the safety of navigation, which translates into the implementation of more than 300 expert rules to ensure the control and deconfliction of bathymetric surveys.

However, different concepts of use might require different rules or preferences to be implemented in the deconfliction process. For example some users, with fewer constraints on the selection of shoals, but stronger constraints on the statical robustness of the bathymetric information (digital terrain elevation surfaces for the use of oceanographic modeling) might welcome relaxed deconfliction rules with the potential weighting of the prioritized sources of overlapping surveys [7,31]. It would be relevant to look at the expert needs concerning the deconfliction rules to be implemented in order to adapt the current workflow to these new practices.

Moreover, with an increasing effort being brought to the automation of the overall workflow, the transformation of the hydrographic profession is questionable. While an effort to generate the first iteration of the reference tiles is currently needed, it is also believed that, through subsequent updates, the hydrographers will have to focus on more and more specific technical issues related to their training without being distracted from minor processing tasks; hence generating a virtuous cycle.

The Téthys workflow systematically implements automation techniques and methodological developments that allow it to take advantage of the intelligence of the data. The generation of the surface reference based on the most relevant bathymetric knowledge allows selected information to be effectively and efficiently provided as support for the generation of marine charts. This methodology and its implementation can prepare the French National Hydrographic to meet the challenges of the future as it better manages bathymetric data, makes it more efficiently usable for end-products, and considers the diversity and increasing volume of bathymetric data to be handled in the close future.

The target is to model all the tiles in the French metropolitan EEZ by the end of the first quarter of 2024; at the time of writing this paper, more than 46% of this area is already produced. Furthermore, updating the bathymetric reference navigation surface with the most up-to-date surface and new incoming surveys, is a crucial task that is easily enabled by the Téthys process.

Author Contributions: Conceptualization, R.J., M.F., J.L.D., T.S. and Y.K.; methodology, R.J., M.F., J.L.D., T.S. and Y.K.; software, R.J., M.F., J.L.D. and T.S.; writing—original draft preparation, J.L.D.; writing—review and editing, J.L.D. and T.S.; supervision, Y.K.; project administration, Y.K.; funding acquisition, Y.K. All authors have read and agreed to the published version of the manuscript.

Funding: This project, financed by the Fonds de Transformation de l'Action Publique, was a structuring project for the Shom. It responded to the need to structure bathymetric information by generating a surface from information from sensors and different acquisition methodologies, and capitalizing on decisions made on it.

Institutional Review Board Statement: Not applicable.

Informed Consent Statement: Not applicable.

Data Availability Statement: Data are not available due to commercial and government restrictions.

Acknowledgments: We would like to extend our gratitude to the Bathymetry Department for all their efforts to complete this project and to the company Geofit for the web portal development.

Conflicts of Interest: The authors declare no conflict of interest.

References

1. Rawsthorne, H.M. De la Sonde à la Carte. Une Campagne Hydrographique au XVIIIe Siècle Dans le Golfe de Gascogne: Analyse Historique et Traitement des Données Obtenues. 2020. Available online: https://dumas.ccsd.cnrs.fr/dumas-02966784 (accessed on 19 February 2023).
2. Caprine-Lancre, J. Une entreprise majeure de la cartographie océanogrphique: La carte générale bathymétrique des océans. *CFC* **2005**, *184*, 67–89.
3. Pastol, Y. Use of Airborne LIDAR Bathymetry for Coastal Hydrographic Surveying: The French Experience. *J. Coast. Res.* **2011**, *62*, 6–18. [CrossRef]
4. International Maritime Organization (IMO). International Convention for the Safety of Life At Sea, 1 November 1974, 1184 UNTS 3. Available online: https://www.refworld.org/docid/46920bf32.html (accessed on 19 February 2023).

5. Elmore, P.A.; Steed, C.A. Algorithm Design Study for Bathymetry Fusion-Review of Current State-of-the-Art and Recommended Design Approach. 2008. Available online: https://www.researchgate.net/publication/235157863_Algorithm_Design_Study_for_Bathymetry_Fusion-Review_of_Current_State-of-the-art_and_Recommended_Design_Approach (accessed on 19 February 2023).
6. Calder, B. On the uncertainty of archive hydrographic data sets. *IEEE J. Ocean. Eng.* **2006**, *31*, 249–265. [CrossRef]
7. Stateczny, A.; Bodus-Olkowska, I. Sensor data fusion techniques for environment modelling. *Proc. Int. Radar Symp.* **2015**, *2015*, 1123–1128. [CrossRef]
8. Gomes, L.; Oliveira, P. Bathymetric data fusion: PCA based Interpolation and regularization, sea tests, and implementation. In Proceedings of the Oceans 2008, Quebec City, QC, Canada, 15–18 September 2008. [CrossRef]
9. Biscara, L.; Maspataud, A.; Schmitt, T. Generation of bathymetric digital elevation models along French coasts: Coastal risk assessment. *Hydro Int.* **2016**, *20*, 26–29.
10. Tozer, B.; Sandwell, D.T.; Smith, W.H.F.; Olson, C.; Beale, J.R.; Wessel, P. Global Bathymetry and Topography at 15 Arc Sec: SRTM15+. *Earth Sp. Sci.* **2019**, *6*, 1847–1864. [CrossRef]
11. Weatherall, P.; Marks, K.M.; Jakobsson, M.; Schmitt, T.; Tani, S.; Arndt, J.E.; Rovere, M.; Chayes, D.; Ferrini, V.; Wigley, R. A new digital bathymetric model of the world's oceans. *Earth Sp. Sci.* **2015**, *2*, 331–345. [CrossRef]
12. Schmitt, T.; Schaap, D.; Spoelstra, G.; Loubrieu, B.; Poncelet, C. EMODnet Bathymetry a compilation of bathymetric data in the European waters. In Proceedings of the OCEANS 2019—Marseille, Marseille, France, 17–20 June 2019; pp. 1–7.
13. Eakins, B.W.; Grothe, P.R. Challenges in Building Coastal Digital Elevation Models. *J. Coast. Res.* **2014**, *297*, 942–953. [CrossRef]
14. Masetti, G.; Andersen, O.; Andreasen, N.R.; Christiansen, P.S.; Cole, M.A.; Harris, J.P.; Langdahl, K.; Schwenger, L.M.; Sonne, I.B. Denmark's Depth Model: Compilation of Bathymetric Data within the Danish Waters. *Geomatics* **2022**, *2*, 486–498. [CrossRef]
15. Kastrisios, C.; Calder, B.; Masetti, G.; Holmberg, P. Towards automated validation of charted soundings: Existing tests and limitations. *Geo-Spat. Inf. Sci.* **2019**, *22*, 290–303. [CrossRef]
16. Hare, R.; Eakins, B.W.; Amante, C.J. Modelling bathymetric uncertainty. *Int. Hydrogr. Rev.* **2011**, *6*, 455–462.
17. Smith, S.M. The Navigation Surface: A Multipurpose Bathymetric Database. Ph.D. Dissertation, University of New Hampshire, Durham, NH, USA. Available online: https://www.opennavsurf.org/assets/papers/smith_thesis_navigation_surface.pdf (accessed on 19 February 2023).
18. NOAA Office of Coast Survey, "BlueTopo". 2023. Available online: https://nauticalcharts.noaa.gov/data/bluetopo.html (accessed on 10 January 2023).
19. *Bureau International Hydrographic International Hydrographic Organisation: IHO Transfer Standard for Digital Hydrographic Data (S-57)*, 3.1 ed.; IHO: Monaco, 2000.
20. International Hydrographic Organization, "S-57 Encoding Bulletins". Available online: https://iho.int/en/s-57-encoding-bulletins (accessed on 20 December 2022).
21. Kirkpatrick, D.G.; Seidel, R. On the Shape of a Set of Points in the Plane. *IEEE Trans. Inf. Theory* **1983**, *29*, 551–559. [CrossRef]
22. Avis, D.; Bremner, D.; Seidel, R. How good are convex hull algorithms? *Comput. Geom. Theory Appl.* **1997**, *7*, 265–301. [CrossRef]
23. Le Deunf, J.; Schmitt, T.; Keramoal, Y. A pragmatical algorithm to compute the convex envelope of bathymetric surveys at variable resolutions. *Abstr. ICA* **2021**, *3*, 1–2. [CrossRef]
24. Samet, H. An Overview of Quadtrees, Octrees, and Related Hierarchical Data Structures. In *Theoretical Foundations of Computer Graphics and CAD*; Earnshaw, R.A., Ed.; NATO ASI Series; Springer: Berlin/Heidelberg, Germany, 1988; Volume 40. [CrossRef]
25. Rockafellar, R.T.; Wets, R.J.-B. *Variational Analysis*; Springer: Berlin/Heidelberg, Germany, 2009.
26. Safe Software, "FME ETL". 2022. Available online: https://www.safe.com/ (accessed on 20 December 2022).
27. Ramsay, P.; Blottiere, P.; Brédif, M.; Lemoine, E. pgPointcloud—A PostgreSQL Extension for Storing Point Cloud (LIDAR) Data. 2022. Available online: https://pgpointcloud.github.io/pointcloud/ (accessed on 19 February 2023).
28. PDAL contributors PDAL—Point Data Abstraction Library. 2020. Available online: https://pdal.io/en/stable/ (accessed on 20 December 2022).
29. Teledyne Caris Caris BDB Python API. 2022. Available online: https://www.teledynecaris.com/en/support/caris-python-api/5-x/bdbserver.html (accessed on 20 December 2022).
30. Agnew, D.C. Robert Fitzroy and the myth of the "Marsden Square": Transatlantic rivalries in early marine meteorology. *Notes Rec. R. Soc. J. Hist. Sci.* **2004**, *58*, 21–46. [CrossRef]
31. Petry, F.; Elmore, P.; Yager, R. Combining uncertain information of differing modalities. *Inf. Sci.* **2015**, *322*, 237–256. [CrossRef]

geomatics

MDPI

Article

Multigrid/Multiresolution Interpolation: Reducing Oversmoothing and Other Sampling Effects

Daniel Rodriguez-Perez [1,*,†] and Noela Sanchez-Carnero [2,3,†]

1 Departamento de Física Matemática y de Fluidos, Facultad de Ciencias, UNED, Avda. Esparta s/n, 28232 Las Rozas, Madrid, Spain
2 Centro para el Estudio de Sistemas Marinos (CESIMAR), CCT CONICET-CENPAT, Bv. Almirante Brown 2915, Puerto Madryn U9120ACD, Chubut, Argentina; noelas@gmail.com
3 Grupo de Ocenografia Fisica (GOFUVI), Facultade de Ciencias do Mar, Campus de Vigo, Lagoas-Marcosende, Illa de Toralla s/n, 36331 Vigo, Pontevedra, Spain
* Correspondence: daniel@dfmf.uned.es
† These authors contributed equally to this work.

Abstract: Traditional interpolation methods, such as IDW, kriging, radial basis functions, and regularized splines, are commonly used to generate digital elevation models (DEM). All of these methods have strong statistical and analytical foundations (such as the assumption of randomly distributed data points from a gaussian correlated stochastic surface); however, when data are acquired non-homogeneously (e.g., along transects) all of them show over/under-smoothing of the interpolated surface depending on local point density. As a result, actual information is lost in high point density areas (caused by over-smoothing) or artifacts appear around uneven density areas ("pimple" or "transect" effects). In this paper, we introduce a simple but robust multigrid/multiresolution interpolation (MMI) method which adapts to the spatial resolution available, being an exact interpolator where data exist and a smoothing generalizer where data are missing, but always fulfilling the statistical requirement that surface height mathematical expectation at the proper working resolution equals the mean height of the data at that same scale. The MMI is efficient enough to use *K*-fold cross-validation to estimate local errors. We also introduce a fractal extrapolation that simulates the elevation in data-depleted areas (rendering a visually realistic surface and also realistic error estimations). In this work, MMI is applied to reconstruct a real DEM, thus testing its accuracy and local error estimation capabilities under different sampling strategies (random points and transects). It is also applied to compute the bathymetry of Gulf of San Jorge (Argentina) from multisource data of different origins and sampling qualities. The results show visually realistic surfaces with estimated local validation errors that are within the bounds of direct DEM comparison, in the case of the simulation, and within the 10% of the bathymetric surface typical deviation in the real calculation.

Keywords: multiresolution interpolation; bathymetry; SRTM; Gulf of San Jorge; Patagonia; Argentina; Atlantic Ocean

Citation: Rodriguez-Perez, D.; Sanchez-Carnero, N. Multigrid/Multiresolution Interpolation: Reducing Oversmoothing and Other Sampling Effects. *Geomatics* **2022**, *2*, 236–253. https://doi.org/10.3390/geomatics2030014

Academic Editors: Giuseppe Masetti, Ian Church, Anand Hiroji and Ove Andersen

Received: 25 May 2022
Accepted: 18 June 2022
Published: 22 June 2022

1. Introduction

Digital elevation models (DEM) are important tools to study the Earth surface and model the processes taking place over it; hazard mapping, climate impact studies, geological and environmental modeling, atmospheric and marine flow simulations including tide prediction, are just a few of their current applications [1–4]. A grid DEM represents the continuous surface interpolated through (discrete) points where elevation has been measured and recorded, and is usually represented as an image whose pixels contain elevation data. High resolution DEMs (∼1 m) appearing in the late 1990s allowed geomorphological exploration with unprecedented detail, both by visual analysis of shaded DEM (e.g., that provides an easy inspection of features at various scales) [5] and through geomorphological indices quantified from the raster image [6,7]. Finding the best DEM generalization (i.e., interpolation) for the scale of topographical features of interest is a key element for multiscale analysis of structural topographic features [5,8,9].

Assessing the accuracy of DEMs is a pending issue, especially for the submerged part of the Earth, where both density and distribution of acoustic bathymetric measurements [10] and spatial resolution (either of interpolation or of indirect gravimetric inversion) are limited. Furthermore, DEM quality is also affected by characteristics of the surface or terrain roughness, cell size or spatial resolution, and the chosen interpolation method (and decisions made about its parameters) [11,12].

Currently, there are different open access global DEMs of the emerged Earth with moderate resolution such as the shuttle radar telemetry model (SRTM, 1 arc second, approximately 30 m horizontal, and 16 m vertical resolution) [13], the ASTER global DEM (GDEM v3, 2.4 arc seconds, approximately 90 m horizontal, and 12 m vertical resolution) [14,15], the Japan Aerospace Exploration Agency (JAXA) AW3D high-resolution global digital surface model (5 m horizontal and 6.5 m vertical resolution) [16], and the ICESat GLAH14 (6 m horizontal and 15 cm vertical resolution) [17,18].

Mapping the submerged bottom of the seas and oceans has required more work. The best known example of open-source bathymetric DEM is the General Bathymetric Chart of the Oceans (GEBCO) [19,20]. Elaborating this DEM involves cleaning and harmonizing data sources and then interpolating them into a surface. Often, this is an iterative process as source data cleaning (and, sometimes, harmonization) cannot be done without an estimated DEM. The acquisition of acoustic data over large areas is very expensive (for a given spatial resolution, it grows with the square of the area), so crowdsourcing strategies are being used to build large databases of bathymetric information [21], being GEBCO one of the most successful ones in terms of integration from multiple sources.

Interpolation methods can be grossly grouped into deterministic, geostatistical and machine learning methods (see the reviews [22–24] for more details):

- Deterministic interpolation methods include nearest (natural) neighbour (NN) [25], inverse distance weighting (IDW) [26], or trend surface mapping (TS) [27]. These methods often work better with homogeneous distributions of data points. There are also models, as ANUDEM a.k.a. ArcGIS TOPO2GRID [28] that are designed to interpolate data along curves (e.g., isolines or river basins).
- Geostatistical interpolation is commonly known as kriging which estimates elevation using the best linear unbiased predictor, under the assumption of certain stationarity assumptions [29,30]. There are many variants that overcome some limitations about those statistical assumptions (such as indicator kriging), or improve prediction based on co-variables (co-kriging).
- Machine learning interpolation methods apply interpolation/classification methods to group "likewise" measurements thus enhancing their efficiency by using previous results. Despite the widespread use of machine learning, its use applied to spatial data is still a field of research; dealing with spatial heterogeneity and the problem of scale are areas in which these techniques can excel (see [31,32]). These methods are also showing their great potential when dealing with multi-source multi-quality data [33].

Interpolated DEMs often present "pimple" artifacts. These are typical of exact interpolation methods, where they appear around sampling points (quite common in IDW), but also appear in approximate (e.g., geostatistical) methods, and are usually removed by filtering the resulting DEM or by increasing the search window. This may cause; however, oversmoothing if the estimated correlation length is larger than the details available in particular areas with higher sampling density; this has been addressed by variance correction methods [34,35]. Another common artifact in DEMs are "transect" artifacts, very common in bathymetric DEMs, that appear where data density is higher (along transects) in contrast with the rest of the raster which is generalized. Some statistical resampling methods have also been devised to address this problem [36,37]. The non-uniform sampling of terrain data can be also caused by selection bias in topographic data (e.g., limited to easily accessible areas), leading to scarcely sampled areas compared with other highly sampled ones.

High accuracy surface modeling methods have been proposed which attempt to overcome this limitation by imposing differential geometry constraints that preserve the expected topographical continuity [38] or introducing pre-interpolated features (e.g., isolines) in the interpolation [36]. Of course, the alternative is to increase sampling effort in undersampled areas; however, this is not always feasible.

The spatial resolution required from a DEM depends largely on the focus of our study interest. For example, a continental DEM or an ocean-wide bathymetry do not require resolving details smaller than several kilometers. On the other hand, the study of coastal tidal dynamics or coastal geomorphometry, or lake or water dam bathymetry may require resolving details of tens of meters or even meters [4,39–42]. When dealing with large areas involving continental scale features data size grows rapidly making it almost impossible to efficiently estimate elevation at points where data are not available, hence techniques are required that are able to efficiently handle large data sets. This interest in multiple scales across large geographical areas has led naturally to multiscale algorithms, either to improve computation of traditional geostatistical interpolations [43], to get advantage of wavelet interpolation algorithms [44], to complete information (especially in bathymetries) by "superresolution" (techniques inherited from digital image inpainting) [45,46], to store and get access to scale-dependent information [47], to analyze scale-dependent geomorphological features [5,9], or even to extrapolate the topography to finer resolutions than available from the data in what is called geostatistical simulation [48,49].

Spatial interpolation methods, either multiscale or not, usually make assumptions about the sampling process (e.g., random independent point-wise sampling), surface statistical properties (e.g., gaussian height distribution, functional form of the variogram), neighborhood shape and extension (e.g., triangulation, look-up distance, look-up directions or quadrants), smoothness penalization or other parameters (curvature constraints, wavelet family, etc.). This makes the choice difficult in common working conditions, statistical assumptions difficult to test, and algorithm parameters difficult to adjust, being the "desirable visual aspect" the most used heuristic criterion in choosing the interpolation, and the software availability and computer memory and processing time the other criteria. The latter are very dependent on the number of points to be interpolated, which again calls for efficient multiresolution approaches.

The goal of this article is to describe multigrid/multiresolution interpolation (MMI) based on simple (if not simplistic) hypotheses about the data, and which is able to solve many of the problems other interpolation methods have, while being fast and extensible. For that we will first introduce a top-down multigrid/multiscale method which meets them while making the simplest hypotheses about the input data or about the interpolated surface (Section 2.1). Then we will show how to use it for surface extrapolation (assuming a self-affine multi-fractal terrain model, in Section 2.3), and cross-validation later used for data filtering and outlier detection (Section 2.4). We will apply this algorithm to two case studies in the area of the Gulf of San Jorge (in Argentina's Patagonia, described in Section 1): one based on synthetic data extracted from the SRTM DEM of the coastal area (Section 3.1), and another based on actual multi-source bathymetric data in order to compute the bathymetric surface of the Gulf (Section 3.2). We will discuss our proposal based on these case studies, and on the current bibliography (Section 4) and, finally, draw some conclusions.

2. Method

Mathematically speaking, interpolation means filling in the gaps of our information about a function based on the information we have about that function, especially but not limited to, the values that function takes at some known points. In what follows, we will construct a multigrid/multiresolution interpolation (MMI) method keeping in mind the geometrical relationships, and properties of exactness, regularity and smoothing, and statistical expectation of the methods described in the introduction. We will also focus on surface interpolation, i.e., interpolation of a real valued function f defined on an interval

$I = [a,b] \times [b,c] \in \mathbb{R}^2$; without loss of generality, we will assume that interval to be $I = [0,1] \times [0,1]$.

2.1. Top-Down Multigrid/Multiresolution Algorithm

Although some interpolation methods aim at providing a grand final mathematical formula to approximate function f at any point $x \in I$, often that formula is not used, but an iterative method estimates the value of f at x from its values $f(\boldsymbol{x}_i)$ at the observation points $\boldsymbol{x}_i \in I$. In addition, in practice, we are often interested in obtaining the average value of the function in some neighborhood $B \subset I$ of x, being the precise value at x often inaccessible experimentally. Based on these two practical approximations, we formulate our multigrid method as follows:

1. Start with a partition of I in $2^{n_0} \times 2^{n_0}$ intervals of the form

$$B_{ij} = \left[i \times 2^{-(n+1)}, (i+1) \times 2^{-(n+1)}\right] \times \left[j \times 2^{-(n+1)}, (j+1) \times 2^{-(n+1)}\right]$$

   with $n = n_0 \in \mathbb{N}$ and $i, j = 0, 1, \ldots, 2^n - 1$. Thus the sidelength of each B_{ij} is equal to $2^{-(n+1)}$, being the sidelength of I equal to 1.
2. Chose those B_{ij} such that for some k there is some observation point $\boldsymbol{x}_k \in B_{ij}$. Let us call N_{ij} the number of those observation points inside B_{ij} and estimate the average value of f in B_{ij} to be

$$\hat{f}(B_{ij}) = \langle f(\boldsymbol{x}_k) \rangle_{\boldsymbol{x}_k \in I_{ij}} = \frac{1}{N_{ij}} \sum_{\boldsymbol{x}_k \in B_{ij}} f(\boldsymbol{x}_k) \tag{1}$$

   This means that our estimation of f in B_{ij} is the most likely one (maximum likelihood) given by the arithmetic mean of the N_{ij} measured points inside B_{ij}.
3. Let us now focus on some B^*_{ij} such that there is no $\boldsymbol{x}_k \in B^*_{ij}$. Let us consider its neighbor intervals, of the form $B_{i\pm\{0,1\}\, j\pm\{0,1\}}$, such that the value of $\hat{f}$ could be computed in them; let us denote that set of neighbor intervals $\mathcal{N}_{ij}$. Then, we will interpolate

$$\hat{f}(B^*_{ij}) = \frac{\sum_{B \in \mathcal{N}_{ij}} w_B \hat{f}(B)}{\sum_{B \in \mathcal{N}_{ij}} w_B} \tag{2}$$

   where the w_B are weights assigned to intervals $B \in \mathcal{N}_{ij}$. The simplest weight assignment would be the number of points inside B, that is $w_{B_{kl}} = N_{kl}$, meaning that we take B^*_{ij} as a part of the larger set $\bar{B}_{ij} = B^*_{ij} \cup \left(\cup_{B \in \mathcal{N}_{ij}} B\right)$ and then we estimate $\hat{f}$ as the average of f over the points measure in that enlarged set $\bar{B}_{ij}$. Under this assumption, we can also interpolate the number of expected measurement points in B^*_{ij} (e.g., after a new statistically independent measurement of the function) as

$$N^*_{ij} = \frac{\sum_{B \in \mathcal{N}_{ij}} w_B N_B}{\sum_{B \in \mathcal{N}_{ij}} w_B} \tag{3}$$

   equating $N_{B_{ij}} = N_{ij}$ in subindices notation.

   Remark 1. *For a partition of I with $n > n_0$, the expression "such that the value of $\hat{f}$ could be computed in them" will also include the rough estimation of $\hat{f}$ (and of N^*_B, $B \in \mathcal{N}_{ij}$) from the previous partition $n-1$ given by (4) below.*

4. Now, we will refine the partition of I by defining, for each B_{ij} four subintervals (quadtree structure), $B'_{ij,kl}$ with $k, l = 0, 1$. If our partition of I was made in $2^n \times 2^n$ intervals, then this one will be in $2^{n+1} \times 2^{n+1}$ intervals of the form

$$B'_{ij,kl} = \left[(2i+k) \times 2^{-(n+2)}, (2i+k+1) \times 2^{-(n+2)}\right] \times \left[(2j+l) \times 2^{-(n+2)}, (2j+l+1) \times 2^{-(n+2)}\right]$$

and assign to each of these subintervals the following values of $\hat{f}$ and $N'_{ij,kl}$ (until a better approximation is made)

$$\hat{f}(B'_{ij,kl}) = \hat{f}(B_{ij}) \qquad (4)$$
$$N'_{ij,kl} = \frac{1}{4} N_{ij}$$

5. At this point, we have for the partition of I in $2^{n+1} \times 2^{n+1}$ intervals a rough estimation of $\hat{f}$, $N'_{ij,kl}$ in each of its subintervals. Then, we can relabel those $B'_{ij,kl}$ subintervals applying the substitution $(ij, kl) \to (2i+k, 2j+l)$ and go back to step 2 to calculate an improved interpolation on a new $n+1 \to n$ partition in new updated intervals B_{ij} of side-length 2^{-n}.

The multigrid quadtree refinement structure of the algorithm makes it to reach a spatial resolution of r (i.e., r is the sidelength of any of the B_{ij} intervals in the last iteration) in $-\log_2(r) - n_0 + 1$ iterations of the previous 5 steps. We only run through the scales in one direction, top-down, hence the title of this section.

2.2. Some Properties of the Algorithm

Exactness: The method is an exact interpolator meaning that, for any partition of I in $2^n \times 2^n$ subintervals, the interpolated $\hat{f}(B_{ij})$ is the mean of observed values of f at points within $B_{ij} \subset I$, in particular for B_{ij} containing one single point (that is the usual meaning of exact interpolation method).

Smoothing: Smoothing of the surface is done during the down-scaling process, applying a nearest neighbors weighted averaging (2) and (3). The neighborhood can be extended to only first-neighbors or to second-neighbors or can be weighted unevenly (e.g., assigning 0.614 weight to second neighbors, assuming octogonal symmetry). In order to get smoother surfaces, the application of Equation (1) can be stopped at some resolution n_s, applying from there on only the generalization operation; then, the method will not be exact at the highest resolution (i.e., pointwise).

Statistical expectation: At every resolution level n, pixels containing data points are asigned the average value of elevation, which is an unbiased estimator of the mean. However, pixels not containing data points are estimated from their surrounding pixels either at that resolution, n, if they contain data points, or at the previous resolution, $n-1$, if they do not. Equations (2) and (3), when used to estimate $\hat{f}(B^*_{ij})$ and N^*_{ij} using as w_B the N_{ij} known up to that level, operate as unbiased estimators acting on unbiased estimations, and then will provide the unbiased expected value of $f(B_{ij})$ when averaged over all possible data samplings. As for the case of ordinary kriging, the underlying hypothesis is that f is "locally constant", hence the neighborhood averaging.

Sensitivity to outliers: As long as the method is based on data averages (or estimated averages), outliers will have their effect on the results. They cannot be safely removed unless strong statistical assumptions (for instance, based on asymptotic standard error of the mean) are made scale-wide, because the same error correction should be applied at all scales. This will be assessed using K-fold cross-validation (see Section 2.4 below).

2.3. Fractal Extrapolation

Geological surfaces, and particularly bathymetric surfaces, are known to evolve through some of these scale-independent transformations and have often been characterized as self-affine fractals [50] or multifractals [51–54] whose Hurst exponent or multifractality spectrum can be related to their geophysical evolution [6,53].

The well known "middle point displacement" method [55] has been used to construct visually realistic ladscape surfaces, and it applies a simple rule to succesively refine a triangulated surface (with some degree of randomness). Although there are variants to this method (among others, to generate multifractal surfaces [56]), the key idea is to make a refinement of the triangulated surface by inserting a new point inside each of its faces (e.g., at the center of the triangles) and assigning to it a height equal to some average of the previous triangle vertices heights plus a randomly distributed zero-mean displacement with variance σ^2 proportional to L^{2H}, being L the side-length of the triangle. The new points, once included in the triangulation, multiply the number of triangles by 3, and the new triangulated surface is transformed by applying the same rule until the required spatial resolution (defined by the triangle side-length L) is achieved.

Given the similarities of this "middle point displacement" construction with our interpolation method, we will adopt it to modify Equation (2) in order to allow for a fractal simulation (or extrapolation) of $\hat{f}$ in those intervals B^*_{ij} without actual measurements $\boldsymbol{x}_k$. So we will just estimate

$$\hat{f}(B^*_{ij}) = \frac{\sum_{B\in N_{ij}} w_B \hat{f}(B)}{\sum_{B\in N_{ij}} w_B} + \frac{1}{\sqrt{12}} s_n \times \eta$$

where η is a uniformly distributed random variable in $[-1,1]$ and s_n is the roughness of the surface (typical deviation) at the scale $L = 2^{-(n+1)}$, given by

$$s_n = \sigma_r \times (L/r)^H$$

where r is the reference resolution (usually, the final interpolated map resolution), σ_r is the estimated roughness at that resolution r (i.e., the root mean square difference between surface heights measured at that resolution) and H is the Hurst exponent.

Usually, H will not be known beforehand, so it can be estimated:

Globally: from the globally mean roughness at the smallest scale (one pixel of the final interpolated map) computed from neighbor height differences Δf between intervals containing observation points. If there are such K pairs of neighboring intervals, then $\sigma_r^2 = s_N^2 = \frac{1}{K}\sum_{k=1}^{K}(\Delta f)^2$. The value of H is estimated from the previous resolution roughness, s_{n-1} which is already known: $H = \log(s_{n-1}/s_N)/\log(2L/r)$. Going global, maximizes the number K, thus the estimation is improved, however local roughness could vary from part to part of the domain.

Locally: in this strategy a value is estimated for σ_r^2 in each interval, using only the neighbor height differences Δf of observation points within that n-th resolution interval (of size L). However, whenever there are no pairs of neighboring points within that interval, σ_r^2 is estimated from the previous resolution (of size $2L$) by the same interpolation method used to estimate $\hat{f}$. This implies that not only $\hat{f}(B_{ij})$ has to be interpolated, but also $\hat{\sigma}_r(B_{ij})$ using the same algorithm.

We will use the local approach in this article.

Remark 2. *Notice that the global estimation of H would play the role of the covariance structure estimation used in ordinary kriging, assuming a power law semivariogram model for the entire area, i.e., assuming a stationary covariance structure. The local approximation would allow for a non-stationary process similar to universal kriging, and also results in multifractal structures. The main difference here is that, as long as possible, the fractal structure is computed as close to the actual scale as possible from measured data, only applying the simulation where necessary, i.e., on intervals with no data for estimation.*

2.4. Surface Validation and Error Estimation

We would like to know how accurate the surface estimation is given a random sample of measurement points $(x_i, f(x_i))$. The common method to assess goodness of fit is validation, that is, using a part of the points not used to fit the function f to compute the distance between the estimated values of $\hat{f}$ at those points and those actually measured values. However, this method only provides a pointwise (at each x_i) or a global (e.g., the mean square error) estimation of error. A bootstrap cross-validation, on the other limit, would repeat the interpolation a large number of times K using each time an independent random sample (extracted "with repetition") of meaurement points, and then estimating the local interpolation error from the distribution of interpolation replicas $\{\hat{f}^{(k)}\}_{k=1...K}$.

In this article, we use a more modest and realizable estimation process, based on K-fold cross-validation. Interpolation will be repeated K times, leaving each time $1/K$-th of the data out. Then, instead of only testing the accuracy of the interpolation with respect to that $1/K$-th of the data, we will estimate the local interpolation standard error $\Delta\hat{f}_{CV}(x)$ from the set of K interpolation replicas $\{\hat{f}^{(k)}\}_{k=1...K}$ as:

$$\Delta\hat{f}^2_{CV}(x) = \sum_{p=1}^{K} \left[\hat{f}^{(p)}(x) - \hat{f}_{CV}(x)\right]^2$$

where

$$\hat{f}_{CV}(x) = \frac{1}{K}\sum_{q=1}^{K} \hat{f}^{(q)}(x)$$

is the mean cross-validation surface.

Remark 3. *Apart from the obvious problem of computing a large number of interpolations posed by bootstrap, the condition of independent random samples poses a problem when measurement data are inherently correlated, as is the case with sampling transects. To address the problem of spatial correlation of points along a transect, we will adopt an "object oriented" K-fold partition of the data. We will subset each transect in smaller sub-transects of equal length (25 km was a practical choice for the case studies below), randomly assigning each of them to one of the K partitions of the data. We will use* $K = 10$*, which is a common choice in the literature [57].*

3. Case Studies

In this section we will apply our interpolation method to reconstruct and assess the quality of two surfaces interpolated from sampled data. First, we will sample data from an area of the SRTM digital elevation model, and test the accuracy of our interpolation both from the sampled data (using the K-fold error estimation) and from comparison with the actual model. Then, we will use bathymetric measurements acquired over an area equivalent in size, and compute the accuracy of our interpolation from those sampled data; in this case, we do not have a more accurate (i.e., computed from more extensive data) bathymetric model than our result, hence the interest of the first one.

Our study cases are located in the Gulf of San Jorge (GSJ) and its adjacent coastal area. The GSJ is is the largest gulf of the Argentinian Patagonian shelf, with an extension of 39,340 km^2 and a mouth of nearly 250 km, located between 45° S (Cape Dos Bahías) and 47° S (Cape Tres Puntas) (Figure 1). This gulf is a semi-open basin mainly covered by silt with coarse granulometric fractions to the north and south ends of the gulf [58,59], that reaches about 100 m of depth in its center, and having in its mouth depths ranging from about 90 m on the north and center, to 50–60 m on the south end, where the basin is demarcated from the adjacent shelf by a pronounced sill. The tidal regime in the GSJ is semidiurnal, with tidal amplitudes ranging between 3–5 m [60,61].

The continental vicinity of the GSJ forms part of the hydrocarbon-producing GSJ basin surrounded by the North Patagonian Massif (north), Deseado Massif (south), and the Andes (west) [62]. These massifs appear in the GSJ as Jurassic rhyolitic volcanic rock outcrops, the larger one located in the northeast (close to Cape Dos Bahías). The GSJ basin plateau is mainly covered by Eocene-Miocene sedimentary rocks of the Sarmiento and Patagonia Formations [63], as well as Quaternary fluvio-glacial deposits ("Rodados Patagónicos"; [64]). This plateau reaches the coast as cliffs or gravel/sand beach-ridges [60].

The GSJ is a very interesting and complex case of management since several interests coexist in it [65]. On the one hand, GSJ is one of the most relevant areas of Argentina coast in terms of biodiversity and productivity with relevant areas for marine conservation because of the presence of reproductive aggregations and foraging grounds of many marine birds and mammals. Moreover, it houses major fisheries targeting valuable shrimp, hake, scallops and king crab stocks [66,67]. On the other hand, its hydrocarbon-producing geology makes it ground of offshore oil platforms [62]. Since each of these processes and activities (oceanographic, fisheries, oil platforms, etc.) extend beyond the limits of the Gulf, we have included in our study the adjacent areas, limited to the north by 44°20′ S (Cabo Raso), south by 48°05′ S (Punta Buque) and east meridian 64° W (Figure 1).

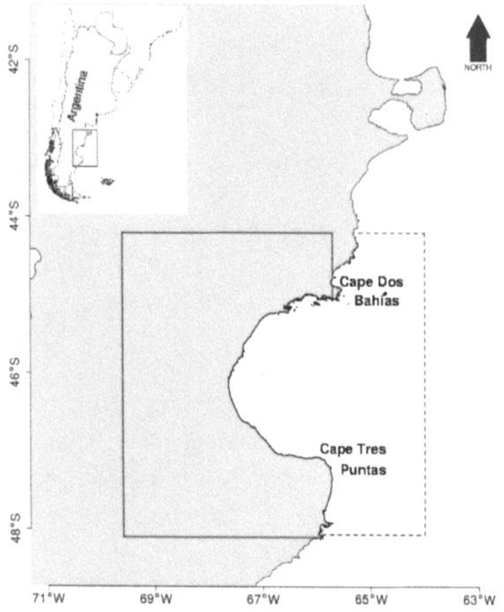

Figure 1. The area of Gulf of San Jorge with the delimitation of the land and ocean regions where the MMI algorithm has been tested.

3.1. *SRTM Digital Elevation Model Sample Reconstruction*

We selected the area between 69°6 and 65°7 W and between 48°1 and 44°2 S shown in Figure 1 (solid line rectangle) for our experiments. The SRTM30 tiles corresponding to this area were merged, resampled and reprojected onto a 90 m UTM grid (zone 19 S); the area includes a total surface of 84,400 km^2 in the emerged zone. Data samples were extracted using two different sampling strategies:

1. random point subsampling;
2. transect subsampling with 25 km long straight parallel transects.

Sampling density, that is the fraction of land points of the grid included in these samples was set to $p = 2^{-n}$, with $n = 4,5,6,7,8$ (that is, from $p \simeq 0.004$ to 0.063). From those samples, a digital elevation model was interpolated with and without fractal extrapolation. For every sampling strategy and density, the average interpolation bias

$$\Delta\hat{f} = \langle \hat{f}_{\rm CV} - f \rangle$$

root mean square error

$$\Delta\hat{f}_{\rm rms} = \sqrt{\langle (\Delta\hat{f})^2 \rangle}$$

the 50% and 90% interquantile ranges of $\Delta\hat{f}$, denoted $IQ_{50\%}\Delta\hat{f}$ and $IQ_{90\%}\Delta\hat{f}$, and the correlation coefficient between $\hat{f}_{CV}$ and f, $\mathrm{cor}(\hat{f}_{CV}, f)$, were computed by direct comparison of the estimated $\hat{f}$ with the full SRTM data f. The K-fold cross-validation mean square errors

$$\Delta\hat{f}_{CVrms} = \sqrt{\langle \Delta\hat{f}^2_{CV} \rangle}$$

are also included in Tables 1 and 2. The K-fold cross-validation estimated standard error $\Delta\hat{f}_{CV}(\boldsymbol{x})$, as well as the standard error map of the interpolated surface, from which table values were computed, are shown in Figure 2. The sampling density of the highlighted column, $p = 2^{-6} \simeq 0.0156$, is the closest one to the sampling density of our case in Section 3.2, $p = 0.0181$.

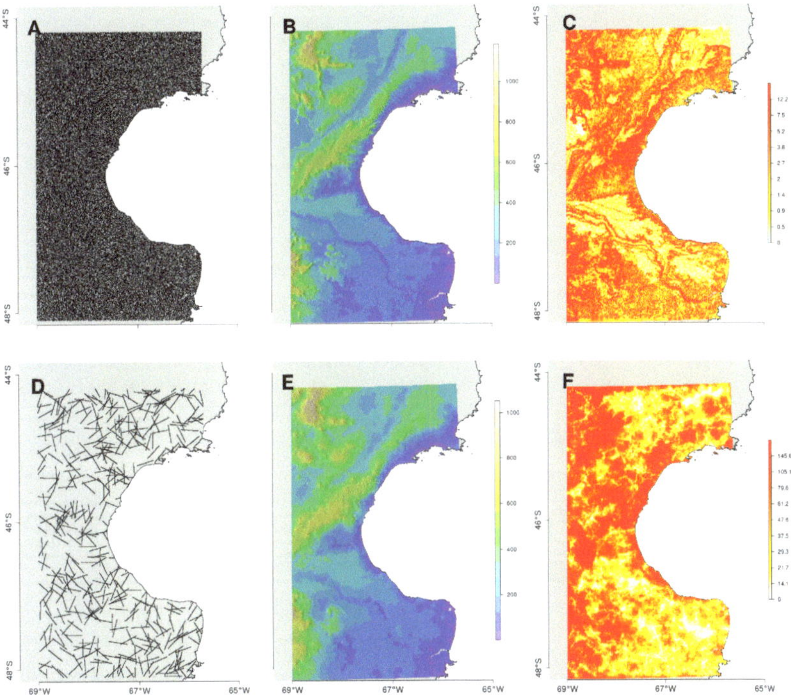

Figure 2. (**A**) Random points used to sample SRTM90 with with $p = 0.0156$; (**B**) Interpolated DEM ($\hat{f}_{CV}$) from point samples; (**C**) K-fold cross-validation standard error $\Delta\hat{f}_{CV}$. (**D–F**) Same meaning, respectively, but using transect sampling.

Table 1. *K*-fold and other statistics of interpolated DEM using simulated data extracted from SRTM at randomly distributed points (p denotes point density per pixel). MMI was applied without and with fractal extrapolation. The SRTM90 column contains an assessment of SRTM resampling error based on the original 30 m resolution SRTM (using *K*-fold cross-validation) for comparison. Values are in meters.

	Simple Interpolation					Fractal Extrapolation					SRTM90
$p =$	2^{-8}	2^{-7}	2^{-6}	2^{-5}	2^{-4}	2^{-8}	2^{-7}	2^{-6}	2^{-5}	2^{-4}	
$\bar{z}$	308.6	308.7	308.7	308.6	308.6	308.6	308.7	308.6	308.6	308.6	308.4
σ_z	184.2	185.1	185.6	185.9	186.2	184.2	185.1	185.7	186.0	186.2	187.2
$\Delta\hat{f}_{\mathrm{CVrms}}$	12.55	10.15	8.23	6.65	5.40	19.33	16.47	15.18	14.58	15.75	2.50
$\overline{\Delta\hat{f}}$	0.044	0.152	0.087	0.045	0.065	0.052	0.182	0.115	0.023	0.088	−0.007
$\Delta\hat{f}_{\mathrm{rms}}$	20.23	16.22	13.11	10.54	8.48	20.85	16.80	13.78	11.44	9.90	0.85
$\mathrm{IQ}_{50\%}\Delta\hat{f}$	8.00	6.53	5.10	4.01	3.13	12.72	10.34	8.58	7.32	7.02	1.25
$\mathrm{IQ}_{90\%}\Delta\hat{f}$	26.45	21.28	17.24	13.86	11.14	37.88	32.33	28.95	26.51	27.35	5.10
$\mathrm{cor}(\hat{f}_{\mathrm{CV}}, f)$	0.9945	0.9965	0.9975	0.9985	0.9990	0.9944	0.9964	0.9973	0.9984	0.9989	1.000

Table 2. *K*-fold and other statistics of interpolated DEM using simulated data extracted from SRTM along random 25 km transects; p denotes the fraction of the raster sampled by the transects. MMI was applied without and with fractal extrapolation. The SRTM90 column shows SRTM resampling error (see Table 1). Values are in meters.

	Simple Interpolation					Fractal Extrapolation					SRTM90
$p =$	2^{-8}	2^{-7}	2^{-6}	2^{-5}	2^{-4}	2^{-8}	2^{-7}	2^{-6}	2^{-5}	2^{-4}	
$\bar{z}$	298.5	311.0	310.9	307.9	307.8	299.4	311.3	310.9	307.9	307.8	308.4
σ_z	140.0	170.6	180.7	178.4	182.6	139.7	170.2	180.8	178.2	182.5	187.2
$\Delta\hat{f}_{\mathrm{CVrms}}$	73.33	53.39	52.70	31.49	23.98	136.55	109.22	88.23	61.90	44.72	2.50
$\overline{\Delta\hat{f}}$	−9.992	2.467	2.355	−0.616	−0.698	−9.168	2.775	2.403	−0.684	−0.708	−0.007
$\mathrm{Var}\Delta\hat{f}$	105.56	79.76	55.92	40.38	28.05	111.59	85.59	60.45	43.97	30.68	0.85
$\mathrm{IQ}_{50\%}\Delta\hat{f}$	49.02	42.17	32.34	20.42	14.99	89.00	78.49	66.26	44.98	31.60	1.25
$\mathrm{IQ}_{90\%}\Delta\hat{f}$	153.37	108.89	106.08	66.91	50.70	247.59	196.91	172.97	126.35	92.25	5.10
$\mathrm{cor}(\hat{f}_{\mathrm{CV}}, f)$	0.8430	0.9113	0.9589	0.9787	0.9899	0.8398	0.9082	0.9582	0.9780	0.9895	1.000

3.2. Gulf of San Jorge Bathymetry Interpolation

Now our area is comprehended between 67°7 and 64°0 W and between 48°1 and 44°2 S as shown in Figure 1 (dashed line rectangle), enclosing a marine area of 85,600 km^2. We used a number of data sources with different spatial sampling strategies (along transects and pointwise), densities, depth reference levels, etc.:

1. Acoustic data from single and split-beam echosounders (SBES): This type of data is distributed in transects, within which there is a very high density of sounding points (depending on the vessel speed and the ping rate, but not greater than one sounding point every ten meters). In addition, the vertical resolution, although dependent on the working frequency, is usually less than 50 cm. In our study case we have several sources of this bathymetric information:
    - The bathymetric data repository published by the National Institute for Fisheries Research and Development (INIDEP) of Argentina, which regularly conducts stock assessment surveys. This repository has a horizontal resolution of one sounding point every 5 m (see details in [68]). In our study area, there were 85085 sounding points, with depths between 11.5 and 123.1 m. These data are distributed in transects located mainly in the northern and southern areas of the GSJ, with less density in the central area.
    - Data from oceanographic campaigns collected in the framework of research project PICT 2016-0218, from the analysis of oceanographic and fishing campaigns carried out by different Argentine intitutions. This database consisted of

147,755 bathymetric points, with depths between 4.2 and 146.7 m. These data are distributed throughout the study area in transects with a mostly NW-SE orientation.
- Data from coastal campaigns. There were 4281 bathymetric points, with depth values between −2 (negative means above low-tide level, that is, the intertidal area) and 71.6 m, all of them acquired with portable echosounders from small vessels. These data are in areas very close to the coast, in the north of the GSJ.

Considering the tidal amplitude ranges in the GSJ, in order to refer all measured depths to a reference low-tide level, a tide correction was applied using the open OSU Tide Prediction Software (OTPS, available from https://www.tpxo.net/otps; access date 17 June 2022) [69].

2. Acoustic data from Multibeam (MBES) and Interferometric Sidescan Sonar (ISSS), which are acoustic sounders that, unlike SBES, provide wide swath coverage, at very high vertical and horizontal resolutions (up to a few centimeters). For our study area, these data come from three acoustic surveys in coastal areas (north of the GSJ), two with MBES and one with ISSS. For this work, the bathymetric surfaces were subsampled onto a 50 m grid. In total, 11,305 bathymetric points were included, with depth values between 5.2 and 121.3 m.
3. Data from nautical charts: the basic source of bathymetric information are always nautical charts, in this case developed and maintained by the Naval Hydrography Services (Servicio de Hidrografía Naval) of Argentina. For our study area, data from six nautical charts were used; one of these charts, covered the entire area, while the other five cover smaller coastal areas, located to the north and west of the gulf, with higher detail. In total, 3522 bathymetric points were used, with depths between 0.3 and 119 m deep.
4. Data from the citicen-science project "Observadores a bordo" (on-board observers, POBCh). Most of the GSJ waters are under the jurisdiction of the province of Chubut, whose Fisheries Secretariat developed the program POBCh for years to control fisheries. In this program, along with fishing data, depth data were taken at those places where fishing sets were made (along with information of date and time). After this database depuration, we used 38,249 bathymetric points in our study area, with depths between 2.8 and 123 m and distributed throughout the entire GSJ except for the SW quadrant, which is under the jurisdiction of another province. Depth data were also corrected using OTPS based on observers annotated coordinates and local time.
5. Coastline. The 0 m isoline of the SRTM30 model was used as the union limit between the emerged and submerged areas. Points were generated along this line, that also includes islands, separated by 20–30 m (a second of arc, corresponding to the SRTM resolution) and with a depth value of 0 m. For the study area, 59,128 points were included from Santa Elena Bay, to the north, to Punta Buque. Coastline is used as a boundary condition and thus not included in the cross-validation process (i.e., it is always included in the interpolation) [36].

In order to harmonize the data, they were subsampled to take one point every 50 m along every transect (to reduce importance bias caused by larger sounding densities) and projected onto a 90 m UTM grid (zone 20 S). Whenever a new data source was projected onto this grid, its depth measurements were corrected to agree on average with the already projected data sources; as a reference, nautic charts were added second, just after the coast line data. The total number of data points within the study area was 248,443.

Remark 4. *Although in some sense this variety of bathymetric sources can be seen as crowdsourced data, all of the data sets were acquired in the context of scientific research programs, and had been previously curated and applied quality tests to remove erroneous data. For example, SBES acoustic data transects were tested for false bottom detections and missing echoes. Similarly, POBCh were checked for the existence of points far off their neighbor depths (usually erroneous manual annotations), and those points were removed from the dataset.*

Outlier Detection

Input data contained a number of points that cross-validation revealed as far-off the mean interpolated surface, sensibly farther than the local standard error $\Delta\hat{f}_{\mathrm{CV}}(\boldsymbol{x}_i)$. To detect them and remove them from the input data we applied the algorithm known as Tukey fences [70] to measurement errors $\hat{f}_{\mathrm{CV}}(\boldsymbol{x}_i) - f(\boldsymbol{x}_i)$. The algorithm consists in calculating the interquartile interval of all these measurement errors and removing those points departing from either interval bound more than k_{Tuck} times its length. That is, only observation points such that

$$\mathrm{Q}_{25\%}\Delta\hat{f}_{\mathrm{CV}} - k_{\mathrm{Tuck}} \times \mathrm{IQ}_{50\%}\Delta\hat{f}_{\mathrm{CV}} \quad < \quad \hat{f}_{\mathrm{CV}}(\boldsymbol{x}_i) - f(\boldsymbol{x}_i) \quad < \quad \mathrm{Q}_{75\%}\Delta\hat{f}_{\mathrm{CV}} + k_{\mathrm{Tuck}} \times \mathrm{IQ}_{50\%}\Delta\hat{f}_{\mathrm{CV}} \tag{5}$$

are kept. According to [70] a value $k_{\mathrm{Tuck}} = 1.5$ does detect outliers, and a value of $k_{\mathrm{Tuck}} = 3$ detects "far off" points; we have used $k_{\mathrm{Tuck}} = 2$ here. We also removed points where $\Delta\hat{f}_{\mathrm{CV}}(\boldsymbol{x}_i)/\hat{f}_{\mathrm{CV}}(\boldsymbol{x}_i) > 0.5$, that is, the cross-validation relative standard error was above 50%; those points did not clearly contribute any information to the interpolation. In total 18,080 points were removed based on these criteria from interpolation in the study area.

After this, another interpolation was carried out again giving the results summarized in Figure 3 and Table 3.

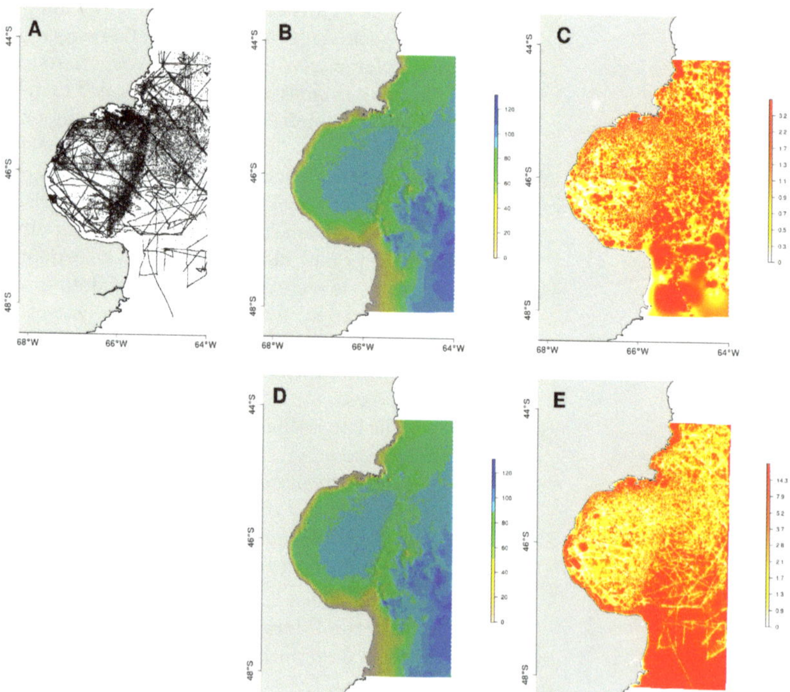

Figure 3. (**A**) Bathymetric acoustic sounding points and transects in the Gulf of San Jorge; (**B**) MMI interpolated bathymetry ($\hat{f}_{CV}$); (**C**) Cross-validation local standard error $\Delta\hat{f}_{\mathrm{CV}}$. (**D**) and (**E**) have, respectively, the same meaning but including fractal extrapolation in the algorithm.

Table 3. Statistics of interpolated bathymetry with real data from the Gulf of San Jorge using the MMI interpolation without and with fractal extrapolation. Values are in meters.

$p = 0.0181$	Simple Interpolation	Fractal Extrapolation
$\bar{z}$	81.32	81.24
σ_z	24.06	24.20
$\Delta\hat{f}_{\mathrm{CVrms}}$	2.02	9.65
$\mathrm{IQ}_{50\%}\Delta\hat{f}_{\mathrm{CV}}$	1.28	4.77
$\mathrm{IQ}_{90\%}\Delta\hat{f}_{\mathrm{CV}}$	4.08	22.87

4. Discussion

Above, we presented and tested a multigrid/multiresolution interpolation (MMI) method with four good qualities: fast, with relatively low RAM requirements (in its simplest version), extensible, based on the fewest possible statistical hypotheses, and locally exact (i.e., at each pixel scale interpolated values concide with the average measured data).

4.1. Asessment of the Interpolations

The potential of MMI is shown in the study cases above. One (Figure 2), in an emerged topography with very different reliefs: from mountains in the north-west (nearing the Andes mountain range) to the southern plains of Patagonian steppe; in addition, a hilly structure runs almost parallel to the coast, from the city of Comodoro (at the midpoint of the GSJ) to the north which, although not having high altitudes, stands out of the sorrounding plains. The other one (Figure 3), in a submerged area combining sandy (south) and rocky (north) coasts, island chains (north), flat sedimentary bottoms (center), basin delimiting sill (east), etc. In both cases, the interpolated surface follows in the larger scales the topography, but also in the smaller ones, if enough data are available, with no appreciable oversmoothig. This is numerically shown in the close values of mean and standard deviation of the original SRTM and interpolated DEMs, with differences below 3% for the mean, and below 10% for the standard deviation and reasonable sampling density (even for transect sampling).

Regarding the interpolation using SRTM sampled data, statistical analysis shows how interpolation cross-validation errors depend strongly on both sampling density and sampling strategy (see Tables 1 and 2): random sampling gives rms standard errors ranging from about 8.5 m ($p = 0.062$) to 20 m ($p = 0.004$), while transect sampling ranges from 28 m ($p = 0.062$) to 105 m ($p = 0.004$), i.e., 4 to 5 times larger; this shows graphically the loss of accuracy far from the transects. The relationship between cross-validation $\Delta\hat{f}_{\mathrm{CVrms}}$ and standard interpolation error $\Delta\hat{f}_{\mathrm{rms}}$ is approximately linear in this range of sampling densities, with $\Delta\hat{f}_{\mathrm{CVrms}}$ slightly underestimating error computed from direct comparison with the original SRTM; nevertheless $\Delta\hat{f}_{\mathrm{CVrms}}$ lays within the 50% and 90% interquantile errors. The inclusion of fractal extrapolation adds to these errors, as expected, but not an statistically significant amount. We can draw from this in order to analyse the GSJ interpolation (Figure 3 and Table 3).

Visually, the GSJ interpolated surface does not show any marked transect artifacts. However "pimple" effects are slightly visible in that interpolation, especially at points from the POBCh data source, and especially in the northern rocky shores (which are naturally irregular) and in the flat sedimentary plateau; it is remarkable that Tukey fences did not remove these points as outliers thus these "pimples" could be just showing real bottom roughness or the need for more sampling in the voids around them. When fractal extrapolation is applied, both effects are masked to some degree by the artificial fractal roughness. A clear case is observed in front of Cape Tres Puntas, where data is scarce and yet the surface is rough, attenuating also the effect of south-leading oceanographic survey transects (but, in turn, increasing the estimated cross-validation error). On the contrary, in the western part of the Gulf the interpolated surface shows a flat bottom with and without fractal extrapolation; this agrees with the known features of the sedimentary seabed in this zone, also confirmed by the relatively low cross-validation error in that area, although in other areas it could result from lack of data there or nearby.

4.2. Assessment of the Method

The idea of multigrid methods appeared in computational mathematics [71] as a way to speedup the solution of partial differential equations and has been interpreted as a preconditioner of the resulting system of linear equations. This not only makes their resolution faster but also numerically more accurate. That was also the goal of using hierarchical basis and wavelets in interpolation methods [44]. Other multiscale methods, either Laplacian/Gaussian pyramid methods in image processing [72], or other wavelet based methods have either been focused on image information representation or compression or on feature analysis [5,44]. In some sense our MMI could be related to some of them, as it uses multiple scale grids (the quadtree structure) that could be formally related to the simple Haar wavelet basis; however, it is difficult to relate those previous works with the interpolation we perform in this work, with randomly distributed point and transect samples, and mean surface estimation at each resolution.

Our MMI method is more easy to compare with other common interpolation methods such as IDW or kriging. It has in common with them that interpolated values are computed as convex linear combinations (i.e., weighted averages) of measured data, without imposing further conditions on the resulting surface. Contrary to kriging, MMI does not require computation of the semivariogram or the stationarity assumptions, which makes it, on the one hand, a (more) parameter free method and, on the other hand, more adaptable to extended areas with subareas of very different elevation profiles. Like these two methods, it is a convex method: interpolated elevations will be weighted averages of measured ones, thus it cannot predict a crest or a valley unless these features were captured by the sampling of the elevation or bathymetric surface (however, see below further improvements that can be included along these lines). MMI, whose underlying idea is just the simple spatial averaging of measuremens inside a tile, is easy to interpret, at least locally; this it has in common with OK which is the best linear unbiased predictor, that is, an estimator of the expected mean elevation based on correlated nearby measurements. The difference here is that MMI assumes measurements to be reliable and aims at interpolating the surface which contains these points, instead of the surface which is the estimated mean of the stochastic surface to which measured points belong.

However, MMI lacks the predictive capabilities of machine learning methods that can predict based on the geophysical features in the area, and are not limited to linear combinations of observations [73]. Contrary to these methods, it can only detect outliers from an statistical assessment of the interpolated bathymetry as we performed in the GSJ bathymetry, although as with any statistical assessment it is not a risk-free decission. Anyway, performing this statistical assessment of the final interpolated surface, using K-fold cross validation as in this work, has other advantages as the spatialization of the estimated error, which is very important in cases with inhomogeneous surfaces, as in the SRTM simulation, or inhomogeneous sampling, as in the GSJ bathymetry [74]. Other potential weakness is related with using data from different sources, but not taking into account their different levels of accuracy. In our approach we took into account these accuracy levels only regarding the vertical reference in the harmonization step (shifting each data set reference to match, on average, the previously more reliable datasets at their cross points). From there on, we applied the common method of rejecting those points far off the general surface trend [21]. Other approaches such as reweighting the data depending on their distance to the average surface (taking into account, or not, local cross-validation error), would have been against our goal of an interpolation method with the least number of assumptions.

Computationally speaking, MMI has also a number of advantages. First, interpolation time is mostly independent of the number of points as the algorithm runs on the quadtree raster pyramid; hence only final raster size determines that time (roughly multiplying it by 4 with every halving of the pixel size). This also means that it will be advantageous when interpolating a large number of data points such as in our bathymetry example: a 3346 × 4928 raster interpolation of 339,874 bathymetric points (padded to 4096 × 8192 pixels for computation) took on average 17 min on an Intel(R) Core(TM) i7-8750H CPU @ 2.20 GHz; the fractal extrapolation took longer: 120 min. Also, being based on raster local operations, it can be adapted to GPU parallel computation (something we have not addressed in our simulations). The fractal extrapolation extension, is not so time-efficient nor so easy to parallelize in the GPU; first, it involves estimating the (multi) fractal distribution parameters, and after that, it requires the use of random numbers for the simulation (which is an issue that has been addressed in other areas such as Monte Carlo simulations [75], but nevertheless increases the complexity of the GPU operations).

Our fractal extrapolation is based on the widely explored characterization of the Earth topography as multifractal [51,53,54]. It takes especially advantage of transect sampling that has been exploited in the past for fractal characterization [76]. It can be seen as a particular approach to geostatistical simulation, that attempts to include complex fine-scale features into (or onto) coarse resolution DEMs, taking into account larger scale spatial height distribution to estimate smaller ones [48]; our estimation method is parametric as it assumes a fractal model. Keeping surface roughness, even if it is simulated, helps to perform terrain classification and regionalization based on geomorphological features computed usually as focal statistics of elevation distribution [7,9], and then terrain classification based on feature distribution across the study area [39,77]; smooth interpolated areas would appear as unreal separate classes, otherwise. From the most basic interest in DEM assessment, fractal extrapolation provides a more realistic estimation of error: in areas where interpolated DEM is totally determined by distant measurements, error can be underestimated based on error propagation (assuming or not a underlying convex formula and gaussian process), or on cross-validation. However, simulating an stochastic surface with the same properties observed in measured areas, will give a more conservative error estimation. Although our method gets this, it is true that some of the simulated features are too random (due to isotropy) and do not prolong the natural trends observed in the area (see, for example, the southern area in front of Cape Tres Puntas in Figure 3).

Future improvements of the MMI algorithm may include extending the generalization window to perform a least squares approximation of the curved surface, weakening the current assumption of a locally flat surface and allowing the inclusion of anysotropy in the fractal extrapolation. This would render more realistic groove and ridge-like features in continuity with the known elevation data [51].

5. Conclusions

In this article, we introduced a multigrid/multiresolution interpolation (MMI) method. The goal of the method is simplicity, both in implementation and in statistical and other assumptions, and scalability to efficiently interpolate large datasets. The quadtree multigrid raster approach makes the method fast and memory efficient. This allows the use of K-fold cross-validation methods to compute local interpolation standard errors, which not only inform about the interpolation quality, but also, helps assess input data quality using outlier detection; this is important when working with heterogeneous data as in our Gulf of San Jorge bathymetry case study.

The (multi)fractal extrapolation method simulates natural roughness in areas with no data (e.g., between transects). On the one hand, this simulates a roughness with the same scale and statistical topographical properties observed in the data (especially in transect data) and, on the other hand, it provides a more realistic asessment of the DEM K-fold cross-validation uncertainty based on the well established multifractal nature of the Earth relief.

We have applied the MMI to synthetic (SRTM elevation model) and real (Gulf of San Jorge bathymetry) DEM interpolation problems, showing how errors depend on sampling strategy and density, and how K-fold cross-validation does a reasonably good job assessing local and global errors. The results show visually realistic surfaces with varying levels of detail, i.e., no oversmoothing, while also reducing transect and "bump" artifacts to a minimum, across a geomorphologically rich area.

Author Contributions: D.R.-P. developed the mathematical and computational aspects of the article. N.S.-C. obtained the data and performed initial standardization and cleaning. Both contributed to write and revise the article. All authors have read and agreed to the published version of the manuscript.

Funding: This work was partially supported by the Agencia Nacional de Promoción Científica y Tecnológica (ANPCyT) of Argentina through project PICT 2016-0218.

Institutional Review Board Statement: Not applicable.

Informed Consent Statement: Not applicable.

Data Availability Statement: The code implementing the algorithms described and some sample data can be found in the GitHub public repository https://github.com/daniel-rperez/mrinterp (access date 17 June 2022).

Acknowledgments: The authors would like to thank Jesus San Martin for his suggestions about method validation. The authors would like to also acknowledge the joint project "Fortalecimiento de la Gestión y Protección de la Biodiversidad Costero Marina en Áreas Ecológicas clave y la Aplicación del Enfoque Ecosistémico de la Pesca (EEP)" between United Nations' Food and Agriculture Organization (FAO) and Ministerio de Ambiente y Desarrollo Sostenible de la Nación Argentina, GCP/ARG/025/GFF, in whose framework were tested some of the methods presented in the article.

Conflicts of Interest: The authors declare no conflict of interest.

References

1. Badura, J.; Przybylski, B. Application of digital elevation models to geological and geomorphological studies-some examples. *Przegląd Geol.* **2005**, *53*, 977–983.
2. Ogania, J.; Puno, G.; Alivio, M.; Taylaran, J. Effect of digital elevation model's resolution in producing flood hazard maps. *Glob. J. Environ. Sci. Manag.* **2019**, *5*, 95–106.
3. Bove, G.; Becker, A.; Sweeney, B.; Vousdoukas, M.; Kulp, S. A method for regional estimation of climate change exposure of coastal infrastructure: Case of USVI and the influence of digital elevation models on assessments. *Sci. Total Environ.* **2020**, *710*, 136162. [CrossRef] [PubMed]
4. Green, J.; Pugh, D.T. Bardsey—An island in a strong tidal stream: Underestimating coastal tides due to unresolved topography. *Ocean. Sci.* **2020**, *16*, 1337–1345. [CrossRef]
5. Kalbermatten, M.; Van De Ville, D.; Turberg, P.; Tuia, D.; Joost, S. Multiscale analysis of geomorphological and geological features in high resolution digital elevation models using the wavelet transform. *Geomorphology* **2012**, *138*, 352–363. [CrossRef]
6. Sofia, G. Combining geomorphometry, feature extraction techniques and Earth-surface processes research: The way forward. *Geomorphology* **2020**, *355*, 107055. [CrossRef]
7. Lecours, V.; Dolan, M.F.; Micallef, A.; Lucieer, V.L. A review of marine geomorphometry, the quantitative study of the seafloor. *Hydrol. Earth Syst. Sci.* **2016**, *20*, 3207–3244. [CrossRef]
8. Marceau, D.J.; Hay, G.J. Remote sensing contributions to the scale issue. *Can. J. Remote Sens.* **1999**, *25*, 357–366. [CrossRef]
9. Newman, D.R.; Cockburn, J.M.; Draguţ, L.; Lindsay, J.B. Evaluating Scaling Frameworks for Multiscale Geomorphometric Analysis. *Geomatics* **2022**, *2*, 36–51. [CrossRef]
10. Alcaras, E.; Amoroso, P.P.; Parente, C. The Influence of Interpolated Point Location and Density on 3D Bathymetric Models Generated by Kriging Methods: An Application on the Giglio Island Seabed (Italy). *Geosciences* **2022**, *12*, 62. [CrossRef]
11. Hengl, T. Finding the right pixel size. *Comput. Geosci.* **2006**, *32*, 1283–1298. [CrossRef]
12. Habib, M.; Alzubi, Y.; Malkawi, A.; Awwad, M. Impact of interpolation techniques on the accuracy of large-scale digital elevation model. *Open Geosci.* **2020**, *12*, 190–202. [CrossRef]
13. Farr, T.G.; Rosen, P.A.; Caro, E.; Crippen, R.; Duren, R.; Hensley, S.; Kobrick, M.; Paller, M.; Rodriguez, E.; Roth, L.; et al. The shuttle radar topography mission. *Rev. Geophys.* **2007**, *45*, RG2004. [CrossRef]
14. Abrams, M.; Crippen, R.; Fujisada, H. ASTER global digital elevation model (GDEM) and ASTER global water body dataset (ASTWBD). *Remote Sens.* **2020**, *12*, 1156. [CrossRef]
15. Tachikawa, T.; Hato, M.; Kaku, M.; Iwasaki, A. Characteristics of ASTER GDEM version 2. In Proceedings of the 2011 IEEE International Geoscience and Remote Sensing Symposium, Vancouver, BC, Canada, 24–29 July 2011; pp. 3657–3660.
16. Tadono, T.; Ishida, H.; Oda, F.; Naito, S.; Minakawa, K.; Iwamoto, H. Precise Global DEM Generation by ALOS PRISM. *ISPRS Ann. Photogramm. Remote Sens. Spat. Inf. Sci.* **2014**, *II-4*, 71–76. [CrossRef]

17. Abshire, J.B.; Sun, X.; Riris, H.; Sirota, J.M.; McGarry, J.F.; Palm, S.; Yi, D.; Liiva, P. Geoscience Laser Altimeter System (GLAS) on the ICESat Mission: On-orbit measurement performance. *Geophys. Res. Lett.* **2005**, *32*, L21S02. [CrossRef]
18. Shuman, C.A.; Zwally, H.J.; Schutz, B.E.; Brenner, A.C.; DiMarzio, J.P.; Suchdeo, V.P.; Fricker, H.A. ICESat Antarctic elevation data: Preliminary precision and accuracy assessment. *Geophys. Res. Lett.* **2006**, *33*, L07501. [CrossRef]
19. Hall, J. GEBCO Centennial Special Issue—Charting the secret world of the ocean floor: The GEBCO project 1903–2003. *Mar. Geophys. Res.* **2006**, *27*, 1–5. [CrossRef]
20. Weatherall, P.; Marks, K.M.; Jakobsson, M.; Schmitt, T.; Tani, S.; Arndt, J.E.; Rovere, M.; Chayes, D.; Ferrini, V.; Wigley, R. A new digital bathymetric model of the world's oceans. *Earth Space Sci.* **2015**, *2*, 331–345. [CrossRef]
21. Novaczek, E.; Devillers, R.; Edinger, E. Generating higher resolution regional seafloor maps from crowd-sourced bathymetry. *PLoS ONE* **2019**, *14*, e0216792. [CrossRef]
22. Li, J.; Heap, A.D. *A Review of Spatial Interpolation Methods for Environmental Scientists*; Record 2008/23; Geoscience Australia: Canberra, Australia, 2008; p. 137. Available online: http://www.ga.gov.au/servlet/BigObjFileManager?bigobjid=GA12526 (accessed on 17 June 2022).
23. Li, J.; Heap, A.D. Spatial interpolation methods applied in the environmental sciences: A review. *Environ. Model. Softw.* **2014**, *53*, 173–189. [CrossRef]
24. Jiang, Z. A survey on spatial prediction methods. *IEEE Trans. Knowl. Data Eng.* **2018**, *31*, 1645–1664. [CrossRef]
25. Yanalak, M. Sibson (natural neighbour) and non-Sibsonian interpolation for digital elevation model (DEM). *Surv. Rev.* **2004**, *37*, 360–376. [CrossRef]
26. Shepard, D. A two-dimensional interpolation function for irregularly-spaced data. In Proceedings of the 1968 23rd ACM National Conference, New York, NY, USA, 27–29 August 1968; pp. 517–524.
27. Chorley, R.J.; Haggett, P. Trend-surface mapping in geographical research. *Trans. Inst. Br. Geogr.* **1965**, *37*, 47–67. [CrossRef]
28. Hutchinson, M.F. A new procedure for gridding elevation and stream line data with automatic removal of spurious pits. *J. Hydrol.* **1989**, *106*, 211–232. [CrossRef]
29. Van der Meer, F. Remote-sensing image analysis and geostatistics. *Int. J. Remote Sens.* **2012**, *33*, 5644–5676. [CrossRef]
30. Maroufpoor, S.; Bozorg-Haddad, O.; Chu, X. Chapter 9—Geostatistics: Principles and methods. In *Handbook of Probabilistic Models*; Samui, P., Tien Bui, D., Chakraborty, S., Deo, R.C., Eds.; Butterworth-Heinemann: Cambridge MA, USA, 2020; pp. 229–242. [CrossRef]
31. Kopczewska, K. Spatial machine learning: New opportunities for regional science. *Ann. Reg. Sci.* **2022**, *68*, 713–755. [CrossRef]
32. Nikparvar, B.; Thill, J.C. Machine learning of spatial data. *ISPRS Int. J. Geo-Inf.* **2021**, *10*, 600. [CrossRef]
33. Kamolov, A.A.; Park, S. Prediction of Depth of Seawater Using Fuzzy C-Means Clustering Algorithm of Crowdsourced SONAR Data. *Sustainability* **2021**, *13*, 5823. [CrossRef]
34. Rezaee, H.; Asghari, O.; Yamamoto, J. On the reduction of the ordinary kriging smoothing effect. *J. Min. Environ.* **2011**, *2*, 102–117.
35. Wang, Q.; Xiao, H.; Wu, W.; Su, F.; Zuo, X.; Yao, G.; Zheng, G. Reconstructing High-Precision Coral Reef Geomorphology from Active Remote Sensing Datasets: A Robust Spatial Variability Modified Ordinary Kriging Method. *Remote Sens.* **2022**, *14*, 253. [CrossRef]
36. Sánchez-Carnero, N.; Ace na, S.; Rodríguez-Pérez, D.; Cou nago, E.; Fraile, P.; Freire, J. Fast and low-cost method for VBES bathymetry generation in coastal areas. *Estuarine Coast. Shelf Sci.* **2012**, *114*, 175–182. [CrossRef]
37. Li, Y.; Rendas, M.J. Tuning interpolation methods for environmental uni-dimensional (transect) surveys. In Proceedings of the OCEANS 2015-MTS/IEEE, Washington DC, USA, 19–22 October 2015; pp. 1–8. [CrossRef]
38. Wang, J.; Zhao, M.W.; Jiang, L.; Yang, C.C.; Huang, X.L.; Xu, Y.; Lu, J. A new strategy combined HASM and classical interpolation methods for DEM construction in areas without sufficient terrain data. *J. Mt. Sci.* **2021**, *18*, 2761–2775. [CrossRef]
39. Sánchez-Carnero, N.; Rodríguez-Pérez, D. A sea bottom classification of the Robredo area in the Northern San Jorge Gulf (Argentina). *Geo-Mar. Lett.* **2021**, *41*, 1–14. [CrossRef]
40. Ibrahim, P.O.; Sternberg, H. Bathymetric Survey for Enhancing the Volumetric Capacity of Tagwai Dam in Nigeria via Leapfrogging Approach. *Geomatics* **2021**, *1*, 246–257. [CrossRef]
41. Perivolioti, T.M.; Mouratidis, A.; Terzopoulos, D.; Kalaitzis, P.; Ampatzidis, D.; Tušer, M.; Frouzova, J.; Bobori, D. Production, Validation and Morphometric Analysis of a Digital Terrain Model for Lake Trichonis Using Geospatial Technologies and Hydroacoustics. *ISPRS Int. J. Geo-Inf.* **2021**, *10*, 91. [CrossRef]
42. Liu, K.; Song, C. Modeling lake bathymetry and water storage from DEM data constrained by limited underwater surveys. *J. Hydrol.* **2022**, *604*, 127260. [CrossRef]
43. Tran, T.T. Improving variogram reproduction on dense simulation grids. *Comput. Geosci.* **1994**, *20*, 1161–1168. [CrossRef]
44. Yaou, M.H.; Chang, W.T. Fast surface interpolation using multiresolution wavelet transform. *IEEE Trans. Pattern Anal. Mach. Intell.* **1994**, *16*, 673–688. [CrossRef]
45. Yutani, T.; Yono, O.; Kuwatani, T.; Matsuoka, D.; Kaneko, J.; Hidaka, M.; Kasaya, T.; Kido, Y.; Ishikawa, Y.; Ueki, T.; et al. Super-Resolution and Feature Extraction for Ocean Bathymetric Maps Using Sparse Coding. *Sensors* **2022**, *22*, 3198. [CrossRef]
46. Zhang, Y.; Yu, W. Comparison of DEM Super-Resolution Methods Based on Interpolation and Neural Networks. *Sensors* **2022**, *22*, 745. [CrossRef] [PubMed]
47. Shekhar, P.; Patra, A.; Stefanescu, E.R. Multilevel methods for sparse representation of topographical data. *Procedia Comput. Sci.* **2016**, *80*, 887–896. [CrossRef]

48. Rasera, L.G.; Gravey, M.; Lane, S.N.; Mariethoz, G. Downscaling images with trends using multiple-point statistics simulation: An application to digital elevation models. *Math. Geosci.* **2020**, *52*, 145–187. [CrossRef]
49. Zakeri, F.; Mariethoz, G. A review of geostatistical simulation models applied to satellite remote sensing: Methods and applications. *Remote Sens. Environ.* **2021**, *259*, 112381. [CrossRef]
50. Blondel, P. *Quantitative Analyses of Morphological Data*; Springer Geology; Springer: Cham, Switzerland, 2018; pp. 63–74. [CrossRef]
51. Gagnon, J.S.; Lovejoy, S.; Schertzer, D. Multifractal earth topography. *Nonlinear Process. Geophys.* **2006**, *13*, 541–570. [CrossRef]
52. Henrico, I. Optimal interpolation method to predict the bathymetry of Saldanha Bay. *Trans. GIS* **2021**, *25*, 1991–2009. [CrossRef]
53. Herzfeld, U.C.; Overbeck, C. Analysis and simulation of scale-dependent fractal surfaces with application to seafloor morphology. *Comput. Geosci.* **1999**, *25*, 979–1007. [CrossRef]
54. McClean, C.J.; Evans, I.S. Apparent fractal dimensions from continental scale digital elevation models using variogram methods. *Trans. GIS* **2000**, *4*, 361–378. [CrossRef]
55. Saupe, D. Algorithms for random fractals. In *The Science of Fractal Images*; Springer: Berlin, Germany, 1988; pp. 71–136.
56. Ebert, D.S.; Musgrave, F.K.; Peachey, D.; Perlin, K.; Worley, S. *Texturing & Modeling: A Procedural Approach*; Morgan Kaufmann: San Francisco, CA, USA, 2003.
57. Wadoux, A.M.C.; Heuvelink, G.B.; de Bruin, S.; Brus, D.J. Spatial cross-validation is not the right way to evaluate map accuracy. *Ecol. Model.* **2021**, *457*, 109692. [CrossRef]
58. Fernández, M.; Roux, A.; Fernández, E.; Caló, J.; Marcos, A.; Aldacur, H. Grain-size analysis of surficial sediments from Golfo San Jorge, Argentina. *J. Mar. Biol. Assoc. U. K.* **2003**, *83*, 1193–1197. [CrossRef]
59. Desiage, P.A.; Montero-Serrano, J.C.; St-Onge, G.; Crespi-Abril, A.C.; Giarratano, E.; Gil, M.N.; Haller, M.J. Quantifying sources and transport pathways of surface sediments in the Gulf of San Jorge, central Patagonia (Argentina). *Oceanography* **2018**, *31*, 92–103. [CrossRef]
60. Isla, F.I.; Iantanos, N.; Estrada, E. Playas reflectivas y disipativas macromareales del Golfo San Jorge, Chubut. *Rev. Asoc. Argent. Sedimentol.* **2002**, *9*, 155–164.
61. Carbajal, J.C.; Rivas, A.L.; Chavanne, C. High-frequency frontal displacements south of San Jorge Gulf during a tidal cycle near spring and neap phases: Biological implications between tidal states. *Oceanography* **2018**, *31*, 60–69. [CrossRef]
62. Sylwan, C.A. Geology of the Golfo San Jorge Basin, Argentina. Geología de la Cuenca del Golfo San Jorge, Argentina. *J. Iber. Geol.* **2001**, *27*, 123–158.
63. Cuiti no, J.I.; Scasso, R.A.; Ventura Santos, R.; Mancini, L.H. Sr ages for the Chenque Formation in the Comodoro Rivadavia región (Golfo San Jorge basin, Argentina): Stratigraphic implications. *Lat. Am. J. Sedimentol. Basin Anal.* **2015**, *22*, 13–28.
64. Martinez, O.A.; Kutschker, A. The 'Rodados Patagónicos' (Patagonian shingle formation) of eastern Patagonia: Environmental conditions of gravel sedimentation. *Biol. J. Linn. Soc.* **2011**, *103*, 336–345. [CrossRef]
65. St-Onge, G.; Ferreyra, G.A. Introduction to the Special Issue on the Gulf of San Jorge (Patagonia, Argentina). *Oceanography* **2018**, *31*, 14–15. [CrossRef]
66. Góngora, M.E.; González-Zevallos, D.; Pettovello, A.; Mendía, L. Caracterización de las principales pesquerías del golfo San Jorge Patagonia, Argentina. *Lat. Am. J. Aquat. Res.* **2012**, *40*, 1–11. [CrossRef]
67. De la Garza, J.; Moriondo Danovaro, P.; Fernández, M.; Ravalli, C.; Souto, V.; Waessle, J. An Overview of the Argentine Red Shrimp (Pleoticus muelleri, Decapoda, Solenoceridae) Fishery in Argentina: Biology, Fishing, Management and Ecological Interactions. 2017. Available online: http://hdl.handle.net/1834/15133 (accessed on 17 June 2022).
68. Sonvico, P.; Cascallares, G.; Madirolas, A.; Cabreira, A.; Menna, B.V. *Repositorio de Líneas Batimétricas de las Campa nas de Investigación del INIDEP*; INIDEP Report ASES 053; INIDEP: Mar del Plata, Argentina, 2021.
69. Egbert, G.D.; Erofeeva, S.Y. Efficient inverse modeling of barotropic ocean tides. *J. Atmos. Ocean. Technol.* **2002**, *19*, 183–204. [CrossRef]
70. Tukey, J.W. *Exploratory Data Analysis*; Addison-Wesley Publishing Company: Reading, MA, USA, 1977; Volume 2.
71. Hackbusch, W. *Multi-Grid Methods and Applications*; Springer Science & Business Media: Berlin, Germany, 2013; Volume 4.
72. Adelson, E.; Anderson, C.; Bergen, J.; Burt, P.; Ogden, J. Pyramid Methods in Image Processing. *RCA Eng.* **1983**, *29*.
73. Sekulić, A.; Kilibarda, M.; Heuvelink, G.; Nikolić, M.; Bajat, B. Random forest spatial interpolation. *Remote Sens.* **2020**, *12*, 1687. [CrossRef]
74. Liu, P.; Jin, S.; Wu, Z. Assessment of the Seafloor Topography Accuracy in the Emperor Seamount Chain by Ship-Based Water Depth Data and Satellite-Based Gravity Data. *Sensors* **2022**, *22*, 3189. [CrossRef] [PubMed]
75. Manssen, M.; Weigel, M.; Hartmann, A.K. Random number generators for massively parallel simulations on GPU. *Eur. Phys. J. Spec. Top.* **2012**, *210*, 53–71. [CrossRef]
76. Malinverno, A. Segmentation of topographic profiles of the seafloor based on a self-affine model. *IEEE J. Ocean. Eng.* **1989**, *14*, 348–359. [CrossRef]
77. Wilson, M.F.J.; O'Connell, B.; Brown, C.; Guinan, J.C.; Grehan, A.J. Multiscale Terrain Analysis of Multibeam Bathymetry Data for Habitat Mapping on the Continental Slope. *Mar. Geod.* **2007**, *30*, 3–35. [CrossRef]

geomatics

MDPI

Article

Denmark's Depth Model: Compilation of Bathymetric Data within the Danish Waters

Giuseppe Masetti [1,2,*], Ove Andersen [1], Nicki R. Andreasen [1], Philip S. Christiansen [1], Marcus A. Cole [1], James P. Harris [1], Kasper Langdahl [1], Lasse M. Schwenger [1] and Ian B. Sonne [1]

1 Danish Hydrographic Office, Danish Geodata Agency, 9400 Nørresundby, Denmark
2 Center for Coastal and Ocean Mapping and NOAA-UNH Joint Hydrographic Center, University of New Hampshire, Durham, NH 03824, USA
* Correspondence: gimas@gst.dk

Abstract: Denmark's Depth Model (DDM) is a Digital Bathymetric Model based on hundreds of bathymetric survey datasets and historical sources within the Danish Exclusive Economic Zone. The DDM represents the first publicly released model covering the Danish waters with a grid resolution of 50 m. When modern datasets are not available for a given area, historical sources are used, or, as the last resort, interpolation is applied. The model is generated by averaging depths values from validated sources, thus, not targeted for safety of navigation. The model is available by download from the Danish Geodata Agency website. DDM is also made available by means of Open Geospatial Consortium web services (i.e., Web Map Service). The original datasets—not distributed with the model—are described in the auxiliary layers to provide information about the bathymetric sources used during the compilation.

Keywords: digital bathymetric model; ocean mapping; open geospatial data

Citation: Masetti, G.; Andersen, O.; Andreasen, N.R.; Christiansen, P.S.; Cole, M.A.; Harris, J.P.; Langdahl, K.; Schwenger, L.M.; Sonne, I.B. Denmark's Depth Model: Compilation of Bathymetric Data within the Danish Waters. *Geomatics* **2022**, *2*, 486–498. https://doi.org/10.3390/geomatics2040026

Academic Editor: Guillaume Ramillien

Received: 17 October 2022
Accepted: 31 October 2022
Published: 11 November 2022

1. Introduction

Ocean bathymetry refers to the depth measurements of the seafloor and, thus, represents the underwater equivalent of land topography [1]. Seafloor bathymetry is commonly distributed using a specialized type of digital terrain model called Digital Bathymetric Model (DBM), which is normally formatted as a regular grid and with depth values assigned to the grid cells [2]. A cursory glance at the available global and regional Digital Bathymetric Models (DBMs) may provide the false impression that the seafloor bathymetry of the oceans is largely known at full coverage. This impression is easily confuted by analyzing the content of these models. The General Bathymetric Chart of the Oceans (GEBCO)—a global DBM, with a resolution of 30 arc sec (e.g., about 926 m at the equator) [1]—lacks actual depth measurements for 80 percent of its coverage [3]. Similar considerations apply to other global compilations (e.g., the Global Multi-Resolution Topography (GMRT) [4]), as well as regional DBMs such as the International Bathymetric Chart of the Arctic Ocean (IBCAO) [5] and the European Marine Observation and Data Network (EMODnet) Bathymetry covering all European sea regions [6]. Although incorporating data derived from both single-beam echosounders (SBES) and modern high-resolution multibeam echo sounders (MBES), these models largely rely on interpolation and altimetry-derived data [1]. Altimetry-derived bathymetry is commonly used by global and regional compilations, but only provides a rough estimation of the seafloor, mainly due to upward continuation in deep waters and variations in sediment and crustal structure on shallow continental margins [7–9]. The depths estimated from altimetry have poor accuracy (i.e., a few hundred meters or worse) and quite low resolution, to the point that only very large seafloor features (in the order of a few kilometers) can be resolved [10].

For the subset of depths in the mentioned DBMs based on actual measurements, the density and the accuracy of the 'soundings' (i.e., bathymetric measurements) vary largely, and this heavily impacts the reliability of the estimated depths. Ocean mapping is limited by the intrinsic characteristics of the ocean environment, particularly by the high attenuation of the electromagnetic waves (i.e., multispectral images from satellites, lidar and radar) in water [11–14]. Thus, the sensors that are widely employed for land topography have limited application—often, just a few meters—in ocean mapping [3]. Instead other types of sensors—e.g., lead-lines and acoustical remote sensing such as SBES and MBES—play a critical role [3]. Historical depths are mostly derived from lead-lines; thus, they are sparse and obtained from a minimal seafloor area (i.e., the few-centimeter diameter of the used weight). When compared to lead-lines, a SBES provides depth measurements that are denser and represent the shallowest point of a fairly large area ensonified by the sonar. In fact, although the position of the measured depth is assumed at the nadir of the surveying platform, its actual location can be anywhere within the ensonified area [15]. Unquestionably, both density and resolution are higher than lead-lines and SBES when using a modern MBES that produces a significantly more accurate representation of the seafloor by electronically forming a set of narrow beams (usually, just a few degrees wide) [16,17]. Unfortunately, only a limited portion of the available models are based on soundings collected with a modern MBES [10]. This is mainly because a MBES for deep waters is physically large and heavy, requiring large platforms to be installed, and thus, relatively expensive to operate [10]. Acoustic geophysical methods also have a primary role in mapping shallow waters, but challenges associated with the coastal environment make it one of the most difficult in which to collect soundings [18] (e.g., the spatial and temporal variability of sound speed [19,20]). Furthermore, the collection of high-resolution bathymetry is not only expensive and frequently challenging, but also time-consuming, as it is only able to cover relatively small regions at a time [21]. Based on these considerations, it should not be surprising that the vast majority of the ocean is still inadequately mapped or even totally unexplored, in spite of centuries of ocean mapping efforts [3].

Due to the difficulties of mapping the seafloor through the water column, our knowledge of the topography of the oceans is largely lagging behind land topography [1]. However, the adoption of advanced techniques to improve the compilation of the available sparse soundings into a DBM has proven beneficial to many fields [21,22]. DBMs are commonly used to accurately describe critical boundary conditions for geophysical, geological, biological, and oceanographic systems [1]. Furthermore, DBM-based analysis is applied in several environmental and geological studies, such as the geohazard and geological analysis of morphologies, with increasing requirements of higher resolutions [23–26]. Elevation surface modelling of coastal areas or entire regions is often based on the integration of DBMs with various types of topographic data [27–29]. Although low-resolution DBMs may be used in global geomorphic features studies [30], they have limited applications in geomorphometric analyses (e.g., benthic habitat mapping) [18,31]. Detailed DBMs are essential to delineate coastlines for storm surges and sea level changes [11], and the morphology of the seafloor, controlling and constraining the bottom currents, and thus, global and regional heat transport [32,33]. Similarly, several aspects of marine geosciences (seafloor characterization, sedimentary studies, offshore engineering, etc.) require high-quality DBMs with meaningful associated metadata [34–36]. DBM's metadata and documentation, describing the main characteristics and limitations associated with a released DBM, facilitates researchers in discovering the bathymetry best fitting their specific purposes [32,37].

Since early 2020, the Danish Geodata Agency have made relevant efforts to organize available bathymetric datasets in Danish and Greenlandic waters into a modern geospatial data management system named DYBDB, and elaborate methodologies to compile these data sources into DBMs and other valuable products (e.g., hydrographic survey overviews) [38]. This paper focuses specifically on Denmark's Depth Model (DDM), the first bathymetric product created employing DYBDB. By improving the bathymetric coverage within the Danish Exclusive Economic Zone (EEZ) currently provided by the EMODnet Bathymetry, one of the major motivations for the creation of the DDM has been supporting environmental studies and other research efforts in the North Sea and in the Baltic Sea.

This paper starts by describing the management of the data sources (along with the main elements of DYBDB), then defines the methodological and technical steps underlying the creation of the DDM. Finally, the content of the publicly available DBM layers and services are presented, with the overall intent of facilitating the adoption of the DDM by researchers and other practitioners.

2. Materials and Methods

2.1. Management of Data Sources

DYBDB is a modern hydrographic data management system that has been designed and implemented by the Danish Hydrographic Office, which is a part of the Danish Geodata Agency.

The DYBDB system is based on several automated procedures (written in Python), task management mechanisms (based on the Atlassian's Jira™ issue-tracking product, https://www.atlassian.com/software/jira, accessed on 30 October 2022), and four types of geospatial databases (see Figure 1):

- **Smart DB**: The Survey Metadata and Raw data Tracker (Smart) database is used to manage an extensive collection of survey metadata, as well as for storing information used to track the integrity of the acquired raw data.
- **Point DB**: The Point database primarily contains the point cloud of cleaned soundings collected during the survey. When available in the data input, the soundings removed during the cleaning process are also stored, thus, replicating the original bathymetric content of the acquired raw data.
- **Grid DB**: Specially designed for dense datasets such as the ones collected by modern MBES, the Grid database contains a subset of the cleaned soundings stored in the Point database, at a spatial resolution tailored for nautical chart production.
- **Model DB:** Intermediate products and final DBMs are stored in the Model database.

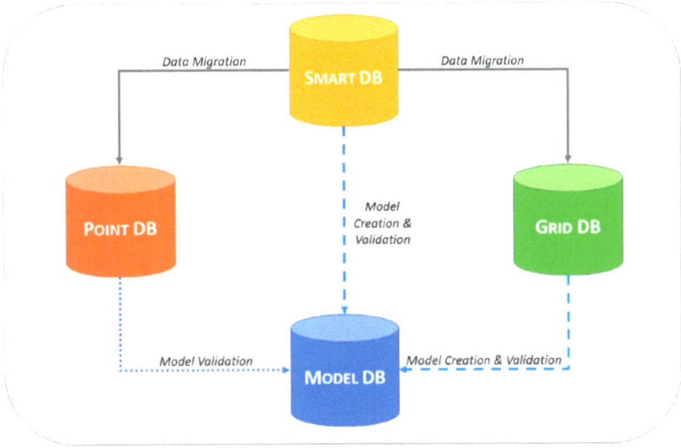

Figure 1. The four types of DYBDB databases (Smart DB, Point DB, Grid DB, and Model DB) and their interactions during key processes. The 'data migration' process (connectors shown in full grey) upload soundings to both Point DB and Grid DB based on the information stored on the Smart DB. The 'model creation' process (in dashed blue) combines soundings stored in Grid DB by retrieving the metadata information from the Smart DB. Once created, the 'model validation' (in dotted and dashed blue) is a semi-automated process that may require access to the point cloud of soundings at full resolution (shown in dotted blue).

The databases use the free and open-source PostgreSQL RDBMS (relational database management system) as backend (https://www.postgresql.org/, accessed on 30 October 2022). Snapshots of the critical content of DYBDB are obtained using the GeoPackage format (https://www.geopackage.org/, accessed on 30 October 2022). All the databases are currently managed through the CARIS' Bathy DataBASE Server™ software, and the CARIS' BASE Editor™ is used as the primary GIS client to access the content of DYBDB (https://www.teledynecaris.com/en/products/bathy-database/, accessed on 30 October 2022).

Since the DYBDB became operational at the beginning of 2020, the Point database and the Grid database have been populated by migrating about 1600 bathymetric datasets (see Figure 1), mainly from hydrographic surveys performed by the Danish Navy, other public agencies, industries, and academia. Most of these datasets have been acquired using SBES and MBES, with sounders generally hull-mounted or installed on a removable pole. Horizontal positioning of the soundings is mainly based on a Global Navigation System (often with corrections to improve accuracy) and, for MBES, an attitude sensor. The latter is required for collecting information on the dynamic movements of the survey platform (i.e., roll, pitch, heave, and yaw) used to spatially orient the acoustic swaths [15,16].

The primary key to uniquely identify a dataset in DYBDB is an encoded textual string named 'Survey ID'. The Survey ID is used not only to retrieve all the soundings belonging to a dataset from the Point DB and the Grid DB, but also to identify a dataset as a contributor to a specific depth value in the Model DB and, finally, in the DDM.

2.2. Compilation Approach

The latest EMODnet Bathymetry (released in December 2020) has a grid resolution of 1/16 arc minute (about 115 m) [39]. As such, to improve the resolution of the publicly available bathymetry within Danish waters, a regularly spaced grid resolution of 50 m was targeted for the DDM. A 50 m resolution was judged to represent a reasonable tradeoff between areas covered with high-resolution surveys (e.g., in the Kattegat area) and regions with only sparse historical soundings (e.g., a large part of the North Sea).

During the processes of model creation and model validation, the DYBDB provides access to datasets and related metadata—specifically, the Smart DB, the Point DB, and the Grid DB—as well as storage for the intermediate products and the finalized DBM in the Model DB (Figure 1). The overall compilation approach is made of the following main steps (Figure 2):

- *Creation/update of the model tiles for datasets in Danish waters.* The source datasets are retrieved from the Grid DB and related metadata from the Smart DB using the *Survey ID*. The sources are gridded by adopting a grid resolution of 50 m and a tiling scheme with a tile area of 1° of latitude by 1° of longitude (Figure 3). The tiles covered by at least one dataset are generated and stored in the Model DB. The bathymetric values are calculated as *representative average depth*, that is, an average of all water depths allocated from the relevant input source to a given grid cell. When multiple datasets overlap, the relevant input source is selected primarily based on the time of data collection. This step is periodically executed to update the tiles in the case of new datasets.
- *Combination of the model tiles into a continuous DBM.* All the populated DDM tiles stored in Model DB are combined into a continuous DYBDB-sources-only DBM.
- *Extension of the continuous DBM with historical soundings.* The DBM calculated in the previous step is extended by combining it with historical soundings available on published nautical products.
- *Interpolation using a Triangulated Irregular Network (TIN) and natural neighbors.* To fill areas with sparse soundings, an interpolated DBM is generated by first creating a Triangulated Irregular Network (TIN) from the extended DBM (generated in the previous step), then using the TIN to interpolate based on the 'natural neighbors' algorithm [40,41].
- *Coverage extraction based on Denmark's EEZ.* The interpolated DBM is updated to limit its coverage from the coastline (generalized at 1:100,000 scale) to the EEZ. The resulting DBM is uploaded to the Model DB.
- *Quality control.* The quality of the DBM resulting from the previous steps is extensively assessed by a team of reviewers. During this iterative process, the reviewers have access to all the direct and indirect DBM sources through Smart DB, Point DB, Grid

DB, and historical data. In case of issues, adjustments to the model may require the (partial or total) re-execution of the previous steps. Only when the outcomes of the quality control are satisfactory is the DBM finalized.

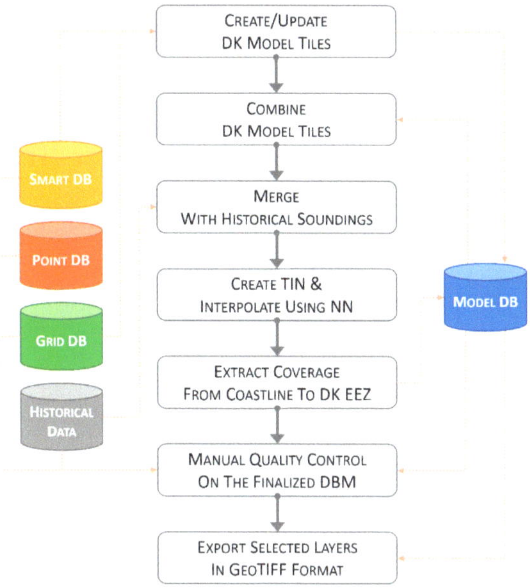

Figure 2. Workflow showing the main steps of the compilation approach (connected using grey arrows). The access (read/write) to DYBDB databases and the retrieval of historical data are shown using orange dotted connectors. Acronyms used in the workflow: DK for Denmark, EEZ for Exclusive Economic Zone, NN for the Natural Neighbor algorithm.

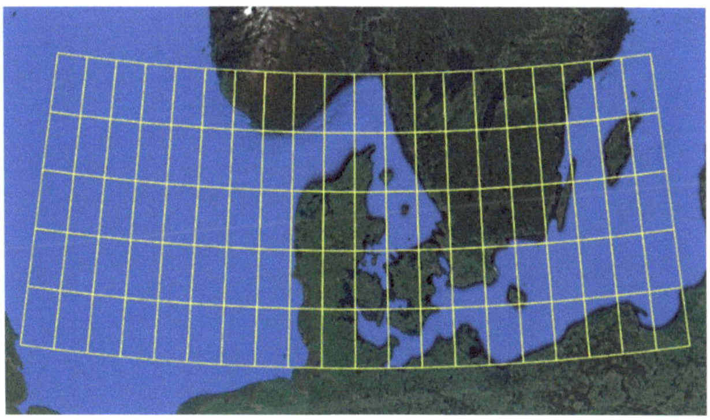

Figure 3. The tiling scheme (in yellow) used to divide the task of compiling the datasets available on the Grid DB. Each tile has an area of 1° of latitude by 1° of longitude. Background from Google Maps' Tile Map Service.

2.3. Model Products

Once the creation and validation processes are completed (following the steps described in the 2.2. *Compilation Approach* section and summarized in Figure 2), the layers listed in Table 1 are exported from the finalized DBM for public release.

Table 1. Layers extracted from the finalized DBM for public release.

Layer (in Danish)	Description
ddm_50m.dybde	The primary layer containing the depth values (in meters).
ddm_50m.kilde	An auxiliary layer providing the source of the depth data for each grid cell. The layer uses the following convention: 1. *DIGI*: The source is a digitalized survey fairsheet. 2. *SB*: The source depths were collected using a SBES. 3. *MB*: The source depths were collected using a MBES. 4. *Historical*: Historical depth values (e.g., lead-line). 5. *Interpolated*: Depth interpolation was applied.
ddm_50m.aar	An auxiliary layer providing the year at which the data collection has ended (only for *DIGI*, *SB* and *MB* dataset types).

The extract layers are projected in Lambert Conformal Conic (LCC)/ETRS89 (EPSG:3034). The vertical datum of the bathymetric layer is a combination of Mean Low Water Spring (MLWS), Lowest Astronomical Tide (LAT) and Dansk Vertikal Reference 1990 (DVR90). The two auxiliary layers (*ddm_50m.kilde* and *ddm_50m.aar*) are used to describe the type and the collection time of the source datasets used to estimate the DDM depths. The original source datasets are not distributed with the DDM. This approach is similar to the one adopted by EMODnet Bathymetry that does not distribute the sources, but provides metadata services (if any) [22].

The output format for the exported layers is GeoTIFF [42]. A *readme* document (in PDF format) with a succinct description on the DDM (i.e., how the model was generated and how to interpret the provided DDM layers) is also a part of the compressed archive containing the DDM release. The DDM layers listed in Table 1 are also made available as Open Geospatial Consortium (OGC) services (i.e., Web Map Service).

3. Results

The official publication of the first release of the DDM happened on 11 November 2022. Both compressed archives containing the material described in *2.3. Model Products* section and information to access the OGC services are available on the Danish Geodata Agency website (https://eng.gst.dk/danish-hydrographic-office/denmark-depth-model, accessed on 30 October 2022).

The released bathymetric layer (Figure 4) covers an area of 232,679 km^2. The largest majority (~97.5%) of the depth values are under 100 m; they present a skewed distribution with a modal depth range between 20 and 25 m and a median value of ~30.5 m (Figure 5).

Based on the *ddm_50m.kilde* auxiliary layer, 18% of the populated grid cells are derived from MBES surveys, and about 75% are derived from interpolation (Figure 6). Based on the *ddm_50m.aar* auxiliary layers, the first MBES-type contribution to the DDM occurred in 1993, and the following years present a significant increase in DDM coverage (Figure 7). The large variability in data density based on the types and years of the DDM sources determined areas with detailed bathymetry derived from MBES surveys (Figure 8), and others that were heavily smoothed because of the interpolation estimating the depth among the sparse soundings (Figure 9).

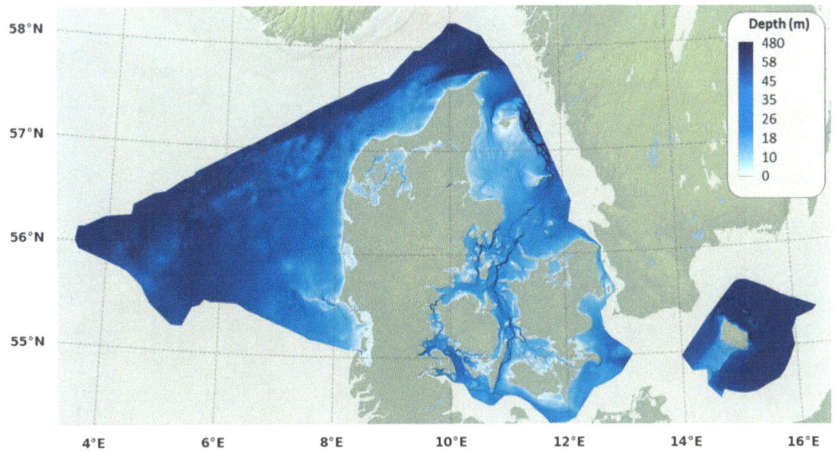

Figure 4. The bathymetric layer of Denmark's Depth Model. The depth values in the color legend are in meters. The Global Multi-Resolution Topography (GMRT) version 4.0 is shown in the background.

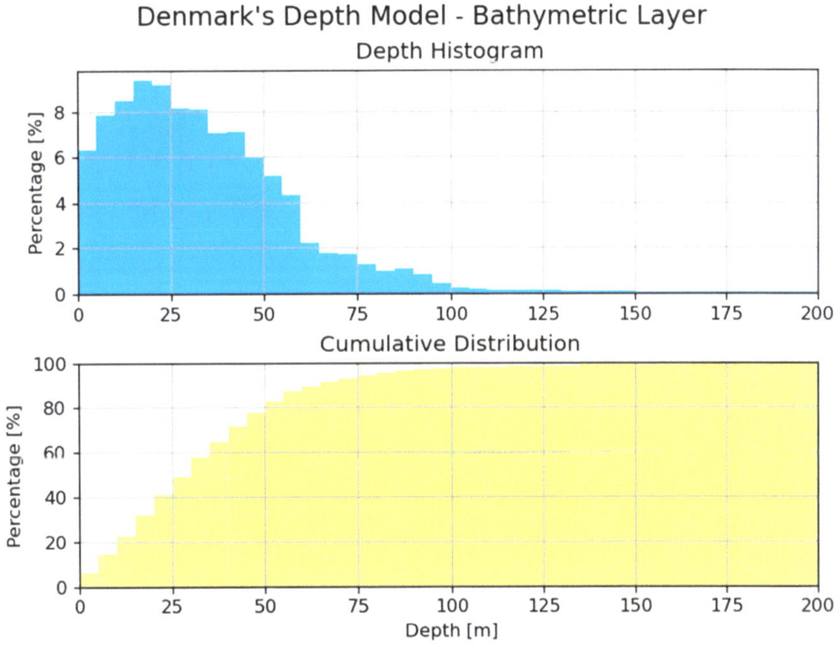

Figure 5. Depth histogram (upper pane) and related cumulative distribution (lower pane) for Denmark's Depth Model's bathymetric layer. For better visualization, an upper limit of 200 m has been applied to the axis of the depth values.

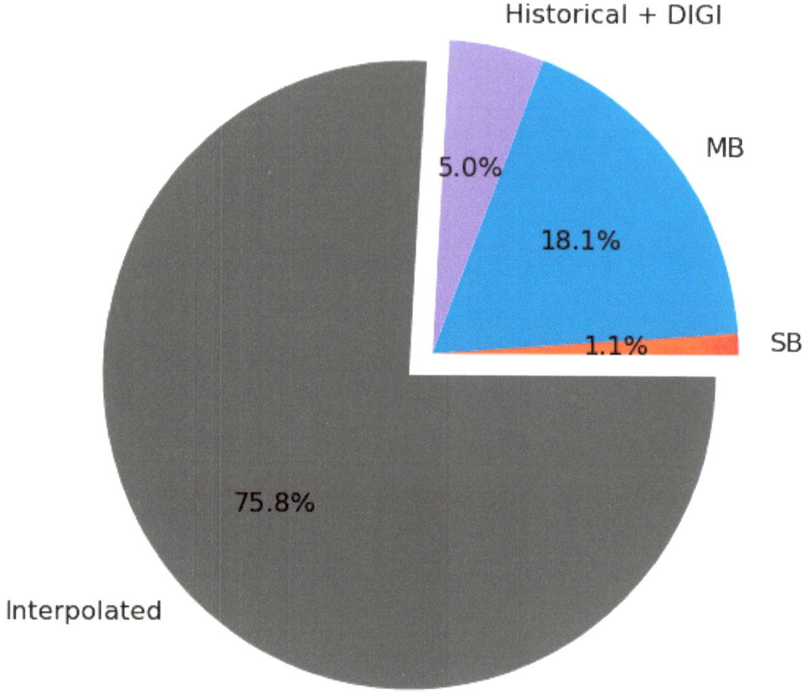

Figure 6. Percentage distribution of the different source types. The labels follow the convention described in Table 1 for *ddm_50m.kilde*.

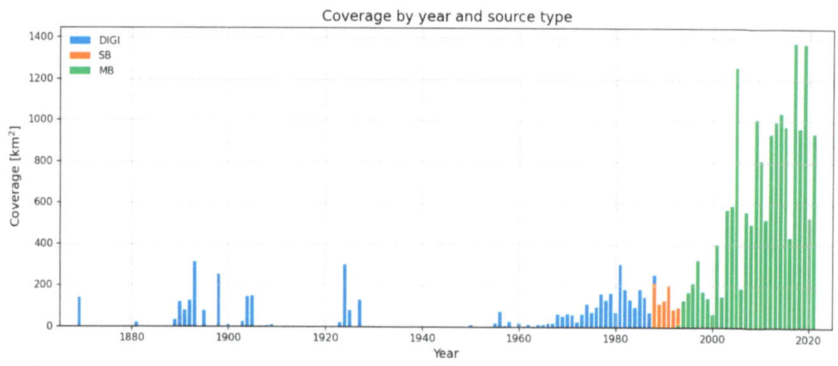

Figure 7. Coverage in km^2 by year and source type. The DDM shows the transition to modern SBES surveys (in orange) in 1988 and the transition to MBES surveys (in green) in 1993.

Figure 8. Perspective view of Denmark's Depth Model at the Great Belt facing north. The area is located between the greater islands, Funen (west) and Zealand (east), and is a heavily trafficked route to the Baltic Sea. The model hill-shading is rendered using a depth exaggeration of 25 times. The maximum model depth in the area is ~70 m.

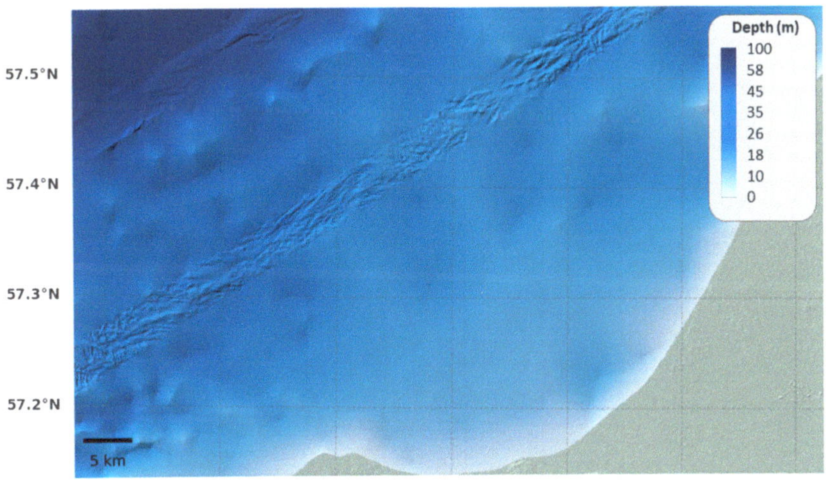

Figure 9. Bathymetry of an area of about 50 km offshore the city of Hirtshals (North Jutland, Denmark). The maximum model depth in the area is ~100 m. The oblique strip with detailed bathymetry is derived from a MBES source. The Global Multi-Resolution Topography (GMRT) version 4.0 is shown in the background.

4. Discussion

Denmark's Depth Model represents the first publicly released model covering the Danish waters with a grid resolution of 50 m. This paper describes the compilation process adopted in the creation of the DDM, as well as its distribution through publicly available products (Figure 4). Both aspects may be of interest for hydrographic offices and other national agencies aiming to actively support research and modeling efforts, given the variety of applications in which DBMs are used. The DDM is generated using an averaging approach, thus, not targeted for safety of navigation. However, several of the steps described in the compilation workflow (Figure 2) can be re-used for future works targeting the development of a navigation surface to streamline the production of nautical charts [43].

The DDM is based on hundreds of bathymetric survey datasets and historical sources within Denmark's EEZ. Unfortunately, less than 20% of the DDM coverage is based on surveys executed with modern SBES and MBES (Figure 6). Significantly increasing this percentage in the coming years is resource-intensive, also because the acoustic swath of MBES is limited by the relatively shallow depths surrounding Denmark (Figure 5). This consideration is one of the main drivers to explore alternative data sources, such as bathymetric lidar and satellite-derived bathymetry—both limited to shallow waters in coastal areas—as well as crowd-sourced bathymetry (CSB). The potential of CSB is large, but its adoption requires practical solutions to overcome a few challenges (i.e., data validation and quality assessment, variable credibility of the collectors) [44].

When modern datasets are not available on a given area covered by the DDM, historical sources are used, or, as the last resort, interpolation is applied. The adopted interpolation approach based on the Natural Neighbor algorithm [40] shows positive results in preserving the details of the areas with dense MBES-type data (Figure 8), as well as in transitioning between areas of wildly different density (Figure 9). However, future works may explore alternative interpolation approaches for introducing further improvements in the DDM [45,46]. Next, releases of the DDM will also likely reduce the interpolated areas, extend the coverage of the inner waters (i.e., fjords, rivers, and lakes), and reduce all the depth values to a common vertical datum (e.g., Mean Sea Level).

The mechanism to compile the hundreds of sources from Grid DB—the "Create/Update DK Model Tiles" step in Figure 2—permits reducing the computation time by requiring updating only the model tiles interested by source changes. More generally, the creation of a robust workflow facilitates the integration of new data sources in the DBM, while preserving a consistent way to present the finalized product. Future work may also explore automated procedures to improve the efficiency of the current quality control of the finalized DBM (Figure 2) [27,47].

DDM has the potential to be beneficial for many scientific applications, from geological studies to oceanography and biology [10,23,48]. Several aspects of marine geosciences—seafloor characterization, sedimentary studies, offshore engineering, etc.—require high-quality DBMs such as the DDM [18,27,35,49]. The metadata and documentation associated with the DDM aims to facilitate its discovery by researchers when searching for the bathymetry best fitting their specific purposes. The downloading services are available on the Danish Geodata Agency website (https://eng.gst.dk/danish-hydrographic-office/denmark-depth-model, accessed on 30 October 2022). The DDM is also made available by means of OGC web services (i.e., Web Map Service).

The original datasets, which are not distributed with the model, are described in the auxiliary layers to provide clear information about the bathymetric sources locally in use by the DBM. Facilitating access to marine data is a critical component of the EU Marine Strategy Framework Directive and the EU Marine Knowledge 2020 agenda, including the already mentioned EMODnet initiative [6,22]. The DDM is also a prospective data source for a future release of the EMODnet Bathymetry. In fact, the EMODnet Bathymetry can receive 'composite grids'—that is, gridded product composed from multiple sources—

as input, by using the SeaDataNet Sextant catalogue service that has been extended for providing details about this type of submission [22].

5. Conclusions

The creation of Denmark's Depth Model (DDM) is based on hundreds of modern datasets (described in the auxiliary layers), historical sources, and interpolation. The resulting DBM represents the first publicly released model covering the Danish Exclusive Economic Zone at a resolution of 50 m.

The current poor knowledge of the ocean seafloor limits our understanding of critical ocean processes providing resources and goods for humanity, controlling the climate, and, more generally, sustaining life on Earth [10]. The DDM improves the bathymetric coverage within the Danish Exclusive Economic Zone (EEZ), which is currently provided by the EMODnet Bathymetry. As such, in times of increasing environmental concerns, the DDM provides a relevant contribution, as described in the United Nations Sustainable Development Goal 14, which aims to "conserve and sustainably use the oceans, seas and marine resources for sustainable development" [50].

Author Contributions: Conceptualization, G.M., O.A., P.S.C., N.R.A. and I.B.S.; software, G.M., O.A., P.S.C. and J.P.H.; validation, M.A.C., K.L. and L.M.S.; writing—original draft preparation, G.M., O.A. and N.R.A.; writing—review and editing, O.A., N.R.A., M.A.C., J.P.H., K.L., L.M.S. and I.B.S. All authors have read and agreed to the published version of the manuscript.

Funding: This research received no external funding.

Institutional Review Board Statement: Not applicable.

Informed Consent Statement: Not applicable.

Data Availability Statement: Current and future release of the model can be retrieved at: https://eng.gst.dk/danish-hydrographic-office/denmark-depth-model (accessed on 30 October 2022).

Acknowledgments: Many past and present colleagues at the Danish Hydrographic Office and the Danish Navy's Hydrographic Service have contributed to this work by providing feedback, data, and endless sources of inspiration. In particular, the authors would like to thank Yvonne Morville Petersen, Torben Weinkouff Rasmussen, Sophie Hohwü-Christensen, and Claus Buus Jensen, for having actively supported the development of the model.

Conflicts of Interest: The authors declare no conflict of interest.

References

1. Weatherall, P.; Marks, K.M.; Jakobsson, M.; Schmitt, T.; Tani, S.; Arndt, J.E.; Rovere, M.; Chayes, D.; Ferrini, V.; Wigley, R. A new digital bathymetric model of the world's oceans. *Earth Space Sci.* **2015**, *2*, 331–345. [CrossRef]
2. Jakobsson, M.; Stranne, C.; O'Regan, M.; Greenwood, S.L.; Gustafsson, B.; Humborg, C.; Weidner, E. Bathymetric properties of the Baltic Sea. *Ocean Sci.* **2019**, *15*, 905–924. [CrossRef]
3. Mayer, L.; Jakobsson, M.; Allen, G.; Dorschel, B.; Falconer, R.; Ferrini, V.; Lamarche, G.; Snaith, H.; Weatherall, P. The Nippon Foundation—GEBCO Seabed 2030 Project: The quest to see the world's oceans completely mapped by 2030. *Geosciences* **2018**, *8*, 63. [CrossRef]
4. Ryan, W.B.F.; Carbotte, S.M.; Coplan, J.O.; O'Hara, S.; Melkonian, A.; Arko, R.; Weissel, R.A.; Ferrini, V.; Goodwillie, A.; Nitsche, F.; et al. Global Multi-resolution topography synthesis. *Geochem. Geophys. Geosystems* **2009**, *10*, 2008GC002332. [CrossRef]
5. Jakobsson, M.; Mayer, L.A.; Bringensparr, C.; Castro, C.F.; Mohammad, R.; Johnson, P.; Ketter, T.; Accettella, D.; Amblas, D.; An, L.; et al. The international bathymetric chart of the Arctic Ocean version 4.0. *Sci. Data* **2020**, *7*, 176. [CrossRef]
6. Schaap, D.M.A.; Schmitt, T. EMODnet Bathymetry—Further developing a high resolution digital bathymetry for European seas. In Proceedings of the EGU General Assembly 2020, Online, 4–8 May 2020; p. 10296. [CrossRef]
7. Sandwell, D.T.; Gille, S.T.; Orcutt, J.; Smith, W.H.F. Bathymetry from space is now possible. *EOS Trans. Am. Geophys. Union* **2003**, *84*, 37–44. [CrossRef]
8. Andersen, O.B.; Knudsen, P.; Berry, P.A.M. The DNSC08GRA global marine gravity field from double retracked satellite altimetry. *J. Geod.* **2010**, *84*, 191–199. [CrossRef]
9. Legeais, J.F.; Ablain, M.; Zawadzki, L.; Zuo, H.; Johannessen, J.A.; Scharffenberg, M.G.; Fenoglio-Marc, L.; Fernandes, M.J.; Andersen, O.B.; Rudenko, S.; et al. An improved and homogeneous altimeter sea level record from the ESA Climate Change Initiative. *Earth Syst. Sci. Data* **2018**, *10*, 281–301. [CrossRef]

geomatics

MDPI

Article

Seafloor and Ocean Crust Structure of the Kerguelen Plateau from Marine Geophysical and Satellite Altimetry Datasets

Polina Lemenkova

Laboratory of Image Synthesis and Analysis, École polytechnique de Bruxelles, Université Libre de Bruxelles (ULB), Building L, Campus du Solbosch, ULB—LISA CP165/57, Avenue Franklin D. Roosevelt 50, 1050 Brussels, Belgium; polina.lemenkova@ulb.be; Tel.: +32-471-86-04-59

Abstract: The volcanic Kerguelen Islands are formed on one of the world's largest submarine plateaus. Located in the remote segment of the southern Indian Ocean close to Antarctica, the Kerguelen Plateau is notable for a complex tectonic origin and geologic formation related to the Cretaceous history of the continents. This is reflected in the varying age of the oceanic crust adjacent to the plateau and the highly heterogeneous bathymetry of the Kerguelen Plateau, with seafloor structure differing for the southern and northern segments. Remote sensing data derived from marine gravity and satellite radar altimetry surveys serve as an important source of information for mapping complex seafloor features. This study incorporates geospatial information from NOAA, EMAG2, WDMAM, ETOPO1, and EGM96 datasets to refine the extent and distribution of the extracted seafloor features. The cartographic joint analysis of topography, magnetic anomalies, tectonic and gravity grids is based on the integrated mapping performed using the Generic Mapping Tools (GMT) programming suite. Mapping of the submerged features (Broken Ridge, Crozet Islands, seafloor fabric, orientation, and frequency of magnetic anomalies) enables analysis of their correspondence with free-air gravity and magnetic anomalies, geodynamic setting, and seabed structure in the southwest Indian Ocean. The results show that integrating the datasets using advanced cartographic scripting language improves identification and visualization of the seabed objects. The results include 11 new maps of the region covering the Kerguelen Plateau and southwest Indian Ocean. This study contributes to increasing the knowledge of the seafloor structure in the French Southern and Antarctic Lands.

Keywords: Antarctic; Southern Ocean; bathymetry; French Southern and Antarctic Lands; cartography; satellite altimetry; marine geophysics; sediments; magnetic anomalies; seafloor

PACS: 91.10.Da; 91.10.Op; 93.85.Pq; 91.10.Fc; 91.10.Jf; 91.10.Vr

MSC: 86A30; 86-05; 86A04; 86A70

JEL Classification: Y91; C83; C88; C90; C00; C60; C61

Citation: Lemenkova, P. Seafloor and Ocean Crust Structure of the Kerguelen Plateau from Marine Geophysical and Satellite Altimetry Datasets. *Geomatics* **2023**, *3*, 393–426. https://doi.org/10.3390/geomatics3030022

Academic Editors: Giuseppe Masetti, Ian Church, Anand Hiroji and Ove Andersen

Received: 11 July 2023
Revised: 6 August 2023
Accepted: 8 August 2023
Published: 10 August 2023

1. Introduction

1.1. Background

In cartography and spatial data processing, the task of plotting maps (also known as mapping layouts) is widely used as a common practice. This involves the visualization and representation of spatially defined objects using cartographic techniques [1–3]. One of the generally accepted methods for mapping and quantitative and qualitative cartographic visualization is implemented through Geographic Information Systems (GIS), using algorithms for raster and vector data processing embedded within these programs [4–6]. However, the high complexity and time-consuming nature of GIS are not conducive to large-scale mapping applications in various Earth science tasks. To address this, programming and scripting algorithms are proposed as an alternative to the GIS-based approach, aiming to reduce data processing complexity in computer graphics and time costs associated with mapping [7].

The application of programming methods in cartography has seen extensive development, resulting in the release of several programs that utilize scripts as mapping tools. Programming-based cartographic approaches can be categorized into three general types. First, there are partial uses of scripts, such as plugins or alternative tools in addition to the existing Graphical User Interface (GUI) in classical modes. Examples include ESRI's ModelBuilder [8], modules in the Geographic Resources Analysis Support System (GRASS GIS), processing script editors in QGIS or ArcGIS [9,10], and the Python-based ERDAS Macro Language (EML) used in Erdas Imagine to create spatial models. Second, there are spatial libraries of programming languages, including selected packages in R and Python specifically designed for satellite image processing and geospatial data analysis [11–14]. Third, there are programs that are completely based on using scripting languages without any Graphical User Interface (GUI), such as the Generic Mapping Tools (GMT). All these examples of using scripts aim to automate cartographic data processing using advanced scripting tools.

The script-based cartographic programs are founded on the principle of utilizing the syntax of a programming language, which includes a number of key commands and recognizable expressions by the system [15]. Following the rules of the embedded language, it becomes possible to compose scripts for data modeling and cartographic visualization. In contrast to traditional GUI-based GIS methods, cartographic scripts do not directly generate maps. However, they provide a series of commands that encompass crucial information about the map's appearance, governing specific features on a plotted cartographic layout [16]. In fact, scripts and programming commands used in cartography define the elements present on the map produced by the script during execution. Specifically, it is possible to define key map concepts such as mathematical definitions of the map (projections, resolution, grid, coordinate systems, extent, and scale), design of symbols and legends (colors and object sizes, palettes for continuous fields, and transparency), and exposition of the elements (overlay, topology, generalization) [17].

1.2. Problem Formulation

The analysis of the seafloor structure and ocean crust concerns mapping, detecting, and recognising the bathymetric structures and variations in geophysical fields. Among these problems, the integrated geophysical interpretation of the satellite altimetry, marine gravimetry, and magnetic anomalies, as well as acoustic and seismic surveys present one of the most active research areas that have attracted research interest in recent decades [18]. Previous studies [19] provided a broad overview of approaches for analysis of the satellite gravimetry data to model the Earth beneath the sea. Marine geophysical and satellite altimetry data provide information on crustal density and processes in the upper mantle [20–22], sediment thickness and basement depth [23]. Moreover, the analysis of the geophysical data enables to reveal key characteristics regarding the lithosphere thickness such as Moho discontinuity and elastic features of the lithosphere [24], viscosity and flexural rigidity. Such data are essential for investigations on the Earth's structure.

In terms of the data-driven analyses, previous works investigating the oceanic seabed can be roughly classified into studies focused on the bathymetric mapping and geophysical analysis. Mapping the seafloor characterizes the bathymetric patterns using data visualization and cartographic methods [25–27]. Marine geophysical methods investigate the geology of the continental margins using methods of deep-sea drilling [28,29], the analysis of features extracted from seismic survey and remote sensing data [30], investigating the structure of the deep ocean basins [31–33], and modelling the mid-ocean ridge system [34–36]. Although traditional methods of bathymetric survey supply novel information on origin, evolution, structure, stratigraphy, and tectonic features of the oceanic crust, they require costly hydrographic equipment such as mutlibeam echo sounder systems [37].

1.3. Related Work

Seafloor bathymetry and geophysical setting of the oceanic crust are two important topics for modelling lithosphere [38]. However, the integrated capture and processing of the multi-source data—such as bathymetry, marine geologic data (development on the Earth's crust including age, spreading rates and symmetry) and marine geophysical data (sediment thickness, gravity and magnetic anomalies)—has always been a challenging task, and few attempts have been made to explore it. Since the variety of data sources is increasing in modern cartography, it results in an exponential growth in the volume of data serving cartographic applications [39]. Therefore, the use of the remote sensing data, satellite altimetry and gravimetry has proven to be effective in geophysical studies. Numerous examples exist in the literature that focus on estimating the geoid values, evaluating gravity anomalies, analysis of the tectonic and crustal structures, glacial and hydrological modelling and habitat mapping.

Satellite altimetry data can be used to jointly represent the patterns of the oceanic currents and to draw conclusions on connectivity between the habitats [40]. For instance, a study by [41] uses the hydrographic and acoustic data for the analysis of the vertical layers of the ocean to identify the community distribution. An example of the tide modelling of bathymetric gradients was presented using the satellite altimetry data [42]. The retreat of glaciers using mass balance measurements was estimated using remote sensing data to analyze spatial distribution of the surface mass balance [43]. Other studies used the Gravity Recovery and Climate Experiment (GRACE) mission for detecting trends and variations in the ice-sheet mass balance by evaluating gravity signals [44]. Furthermore, Mathieu et al. [45] combine the SRTM DEM, remote sensing data and geological sampling to clarify seafloor structures.

These and similar examples illustrate that a combination of bathymetric and geophysical data for the detailed analysis of Earth's crust structure and topography outperforms their use separately. This suggests that seafloor mapping should be based on the complex analysis of the heterogeneous features of the seabed. Such information can be derived from various Earth observation datasets, as pointed out earlier [46–49]. Since such methods can be regarded as feature-level combinations of seafloor features showing the local appearance of prominent seabed elements such as fracture zones, ridges, seamounts, deep-sea trenches, canyons, or rifts, the use of a programming approach to merge different integrated geophysical datasets enables matching multi-source data with varying resolutions and origins. Such data can be used for the detailed analysis of seafloor structure and geodynamic processes [50–52].

1.4. Objection and Motivation

In this study, several bathymetric and marine geophysical datasets were used for the analysis of the Kerguelen Plateau (Figure 1) within the south-western segment of the Indian Ocean to visualize, describe and analyze the structures of the seafloor in the context of the geophysical settings. Ten new maps are presented and described to visualize spatial variations, reveal correlation and matches between the geophysical and topographic setting of the Kerguelen region to continue the existing similar studies [53–55]. Given the remote location of the Kerguelen Archipelago, one of the most isolated places on Earth, and associated difficulties and high cost of geological and geophysical sampling, the integrated use of the satellite-derived data enables to us have better insight into the structure of the Kerguelen Plateau.

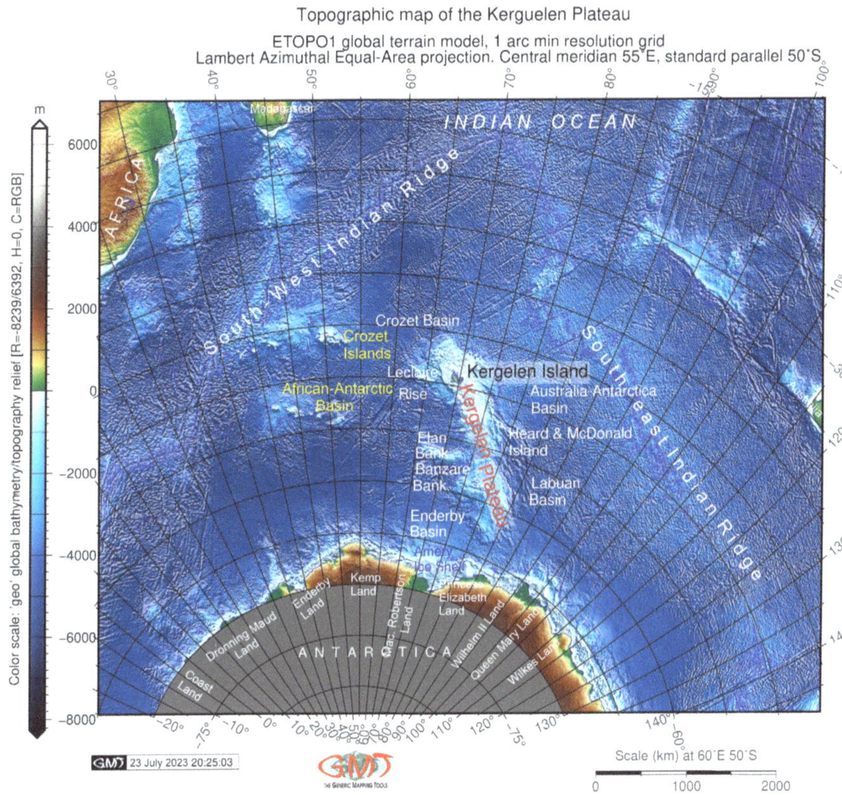

Figure 1. Map of the Kerguelen Plateau region. Mapping: GMT. Map source: author.

2. Study Area

The Kerguelen Plateau presents a large topographic elevation in the south-west Indian Ocean, extending over 6500 km^2 [56], Figure 1. The Kerguelen Archipelago, formed on the plateau, constitutes a shallow submarine plateau. The structure of the Kerguelen Archipelago is asymmetric with notable variations in the two distinct parts: the northern part (Heard, McDonald and Kerguelen Islands) is more shallow (<1000 m) compared to the southern segment (depths of 1500–2000 m) [57]. Morphologically, the plateau is oriented in a NNW direction toward the Antarctic continental margin where it is constrained by Princess Elizabeth Land. On the northwest, the archipelago is limited by the Crozet Archipelago, located 1250 km away [58]. To the north, it is bordered by the African-Antarctic Basins; to the northeast, by the Australian-Antarctic Basin; and to the south, by the Antarctic. Adjacent lands include Enderby Land, Kemp Land, Mac Robertson Land, Princess Elizabeth Land, Wilhelm II Land, and Queen Mary Land, all located in the Antarctic, as shown in Figure 2.

The origin of the Kerguelen Plateau is related to the hotspot associated with the Gondwana breakup in the Cretaceous period (120–110 Ma), and seafloor spreading between India and Antarctica in the south-western Indian Ocean [59]. The initial surface manifestations of the Kerguelen Plume in the southern region began during the Mesozoic era around 120 Ma [60]. Active submarine volcanism occurring on the aseismic ridge, coupled with sedimentation processes since the Cretaceous, led to the creation of the basaltic basement of the Kerguelen region [61]. This volcanic activity persisted within the Kerguelen hotspot until the Tertiary period (<40 Ma) and is evident in the geological and mineralogical composition of the archipelago. The archipelago is covered by up to 85% basalts originating from the mantle plume of the hotspot. These basalts result from the crystallization of raised and cooled basaltic magma [62–64]. Presently, the Kerguelen Plateau stands as one of the largest volcanic plateaus globally, spanning a total length of 2300 km. Remnants of volcanic activity are still visible on the Heard and McDonald Islands [65].

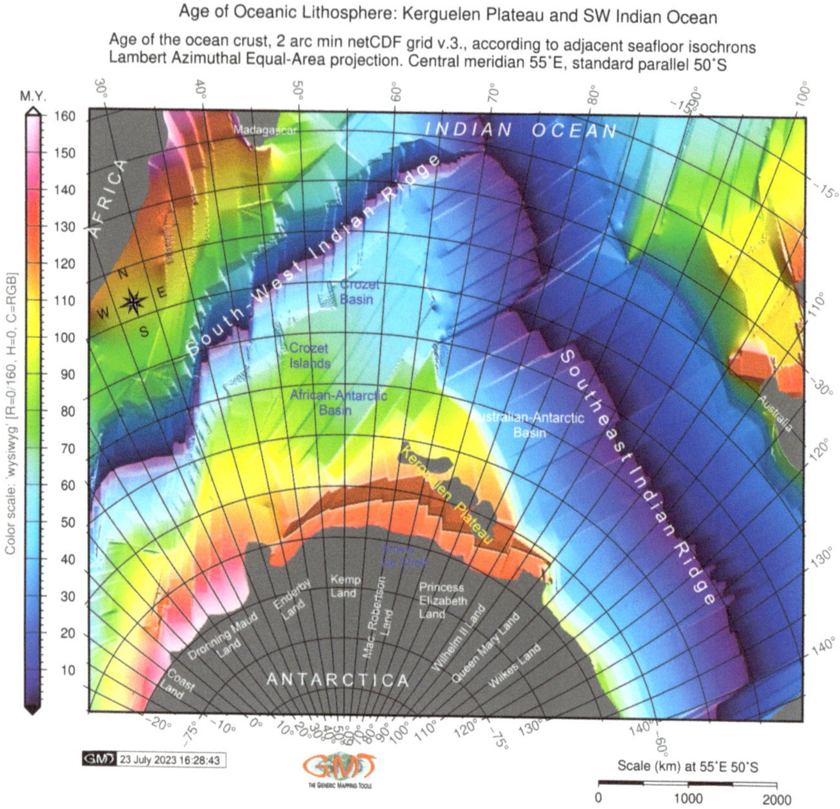

Figure 2. Age of the ocean crust in SW Indian Ocean and the Kerguelen Plateau region and east Antarctic. Mapping software: GMT v. 6-1-1. Map source: author.

The Kerguelen Plateau belongs to the French Southern and Antarctic Lands, known for its protected environment [66–68]. Its high environmental value is explained by the unique wildlife structure, which includes vulnerable species. The remote location of the archipelago—4000 km from South Africa and Australia [69,70]—has created ideal conditions for preserving unique flora and fauna of the Kerguelen Archipelago. The presence of rare Antarctic plant species and high biodiversity is closely linked to the soil developed on the basaltic basement and the specific geological substrate of volcanic origin. Additional factors contributing to this uniqueness include the influence of the Polar climate [71–73]. The distribution of fish communities and phytoplankton is affected by seasonal changes of the Antarctic circumpolar currents [74–81]. Furthermore, the steep bathymetric slopes and the exposure of the Kerguelen plateau amplify the speed and intensity of the ocean currents' circulation [82]. The combination of all these factors makes the ecosystems of the Kerguelen Archipelago unique and deserving of protection as an environmental heritage of the French Southern and Antarctic Lands [83–85].

3. Materials and Methods

All the maps have been plotted using the scripting toolset Generic Mapping Tools version 6.4.0 [86], https://www.generic-mapping-tools.org/. A key aspect of the GMT algorithms is that they consider each cartographic element by its parameters and add new layers on the map regarding the target location, which can be adjusted using refined flag options in special modules. Thus, a modular scripting approach of GMT distinguishes it from GIS.

3.1. Data

The materials used as input cartographic grids include the following datasets. The marine geological data on age, spreading rates, and asymmetry of the ocean crust were derived from the NOAA high-resolution dataset [87], Figure 3. The units of the dataset on spreading asymmetry are expressed as % of crustal accretion of the seafloor, i.e., symmetric spreading results in values of 50% on conjugate ridge flanks. The bathymetric mapping was based on the ETOPO Global Relief Model [88]. The gravity grids were obtained from the EGM-2008 and EGM96 [89]. Gravity data are grounded on the concept of gravity anomaly across different Earth surfaces, which correlates with the interplay between topography, mass distribution within local and regional relief structures, marine gravity values, and the establishment of gravitational potential across various Earth locations.

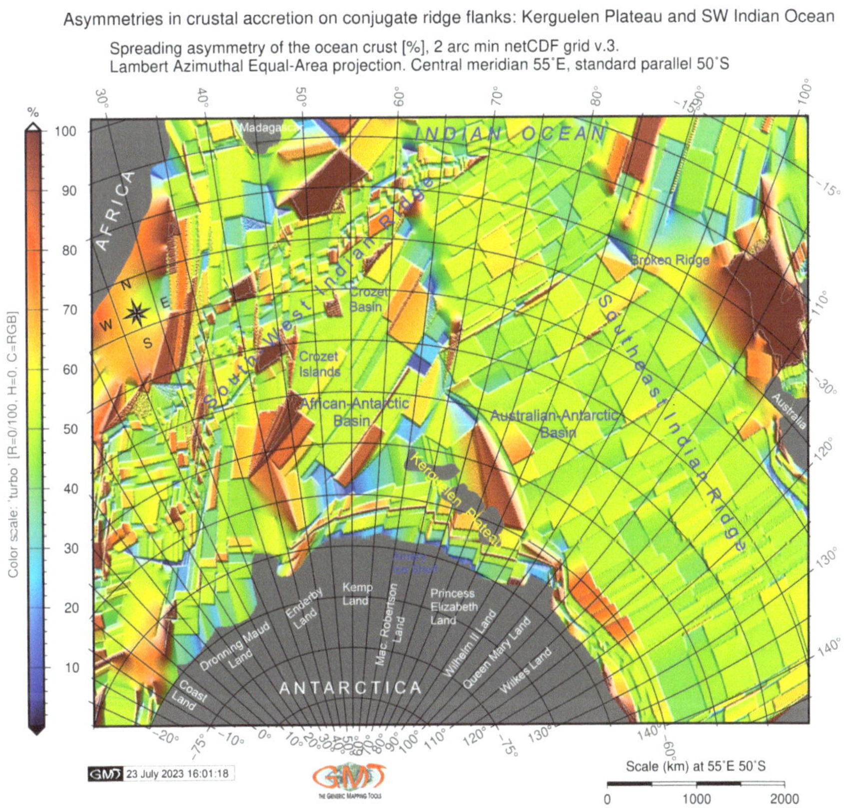

Figure 3. Asymmetry in crustal accretion on conjugate ridge flanks of the ocean crust over the Kerguelen Plateau region, east Antarctic and south-west Indian Ocean. Map source: author.

The free-air and Bouguer gravity anomaly grids are derived from high-resolution grids [90,91]. Specifically, the gravity field varies significantly over the oceans due to density fluctuations. The most pronounced anomalies emerge from fluctuations in density, such as those occurring at the heterogeneous seafloor or at the crust-mantle interface (Moho discontinuity) [92]. Such data facilitate the modelling of gravity fields for the analysis of

Earth's structure. The latest updates in gravity grids enhance the precision of altimetry-derived gravity anomalies. While gravity grids do not directly indicate topography, they offer crucial insights into the Earth's relief, providing an approximate relationship between topographic and geophysical data and illustrating the correspondence of surface gravity values with local geoid undulations [93].

The data on sediment thickness are collected from the GlobSed dataset, which models distribution of the sediment thickness in the World's Oceans [94]. The magnetic data are derived from the two available information sources compiled originally from satellite and marine magnetic measurements: the Earth Magnetic Anomaly Grid (EMAG2) [95], which has a resolution of two arc minutes, and the World Digital Magnetic Anomaly Map (WDMAM) with a resolution of three minutes [96]. The gravity data were derived from the free-air global gravity grids [97].

The analysis of the geodetic data is useful in such complex terrains since it facilitates the identification of major tectonic structures and topographic patterns. In addition, the EGM grids, the identification of lithospheric structure and tectonic blocks can also be performed using satellite gravity data such as World Gravity Model (WGM) and GRACE [98]. Furthermore, the EGM and astrogeodetic vertical deflections are useful for modelling gravity and geopotential differences [99]. Finally, processing terrestrial and satellite geodetic data can be applied to address theoretical geodetic challenges, such as the altimetry–gravimetry boundary-value problem [100]. Other instances of the implications of geodetic datasets encompass the utilization of satellite-derived data from Cryosat-2 and altimetry in conjunction with ship-measured gravity to estimate marine gravity anomalies [101].

3.2. *Methods*

In this study, the cartographic methodology is based on using the Generic Mapping Tools (GMT) programming suite version 6.4.0 [102] by the developed scripting workflow, explained in detail in earlier works [103,104]. The most prominent feature of the GMT is a scripting approach that principally distinguishes it from the conventional software due to the embedded programming language [105]. The traditional GIS-based methods either employ the GUI for mapping with existing standard menu or allow scripting as a complimentary workflow.

In contrast, the GMT is a completely console-based software that operates entirely using scripts. In this way, it captures the data by running a script written using the embedded syntax that operates similarly to programming. A script consists of a sequence of predefined commands with parameters that control the appearance of cartographic elements and features. The most well-known cartographic tools that employ scripts for data processing are GMT [106] and GRASS GIS [107]. Other software allows scripts as an additional functionality alongside the standard GUI, as seen in ArcGIS or QGIS. While the latter is predominantly used for image processing, cartographic, and ecological studies [108–111], GMT finds application in geophysical and geological research [112–114].

3.2.1. Topographic/Bathymetric Mapping

The script for plotting the topographic/bathymetric map is provided in Listing A1. The code's most crucial elements are as follows. The mapping process initiates by selecting the study area from the global ETOPO1 grid using 'grdcut'. Isolines are chosen through 'grdcontour' with a 1,000 m gap. In our case, these lines denote significant locations that offer a high representation estimate for seafloor bathymetry. 'grdimage' is utilized to generate an image plot of the raster grid, colored as defined by the 'makecpt' module, with values -T indicating the extreme (min/max) data for the topographic grid, as extracted by the 'gdalinfo' module. The remaining code details are outlined in Listing A1 along with key comments.

3.2.2. Mapping Seafloor Age, Spreading Rates, Spreading Asymmetry and Age Uncertainty of the Ocean Crust

Mapping the age, age uncertainty, spreading rates, and spreading asymmetries of the Kerguelen Plateau is executed using raster grids with a 2-min resolution, as shown in Figure 4. These grids are reprojected to the Lambert Azimuthal Equal-Area projection utilizing data from major features within the Indian Ocean basin. The 'psxy' module is employed to add vector lines representing mid-ocean ridges, tectonic plates, and lithospheric plates. This provides a depiction of the borders of the Indian, African, and Australian plates, as seen in Figure 4. Spreading half rates of the ocean crust are visually presented using the 2-arc-minute netCDF grid v.3.6 from NOAA. Similarly, tectonic slab contours are added via the 'psxy' module of GMT. Multiple grids are combined and displayed as plotted raster grids to illustrate age-depth relationships in the seafloor around the Kerguelen Plateau and southwest Indian Ocean. The complete scripts are available in Appendix A of this study, with Listing A2 utilized for mapping the age of the oceanic lithosphere over the Kerguelen Plateau, Listing A3 for mapping asymmetries in crustal accretion on conjugate ridge flanks over the Kerguelen Plateau, Listing A4 for visualizing spreading half rates of the oceanic lithosphere in the Kerguelen Plateau and SW Indian Ocean, and Listing A5 for mapping crustal age uncertainty of the oceanic lithosphere over the Kerguelen Plateau, as depicted in Figure 5.

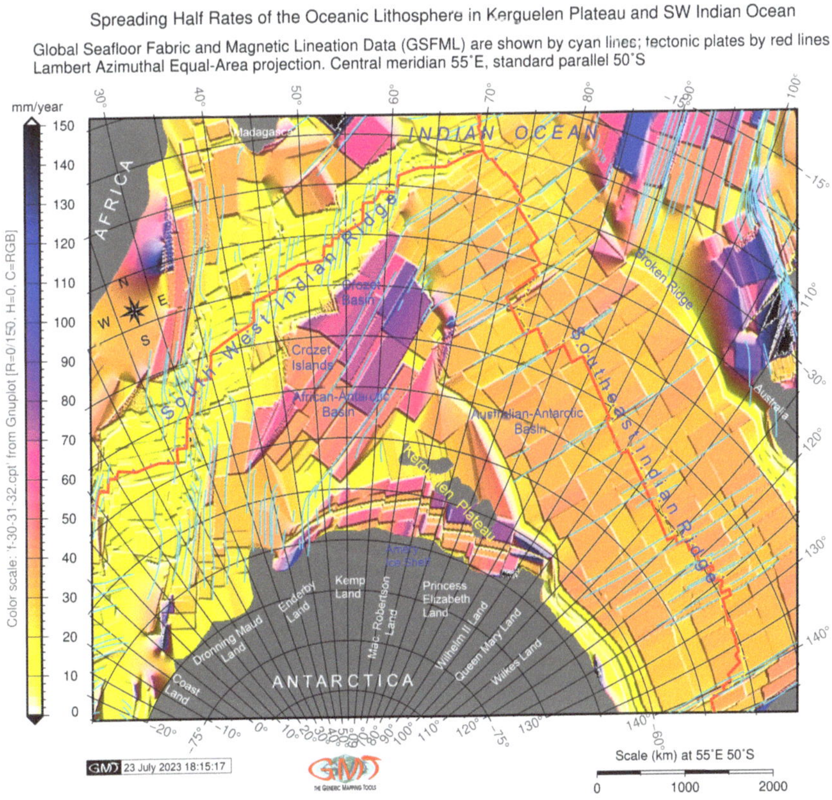

Figure 4. Spreading half rates of the ocean crust over the Kerguelen Plateau region, east Antarctic and south-west Indian Ocean with added GSFML data. Mapping tool: GMT. Map source: author.

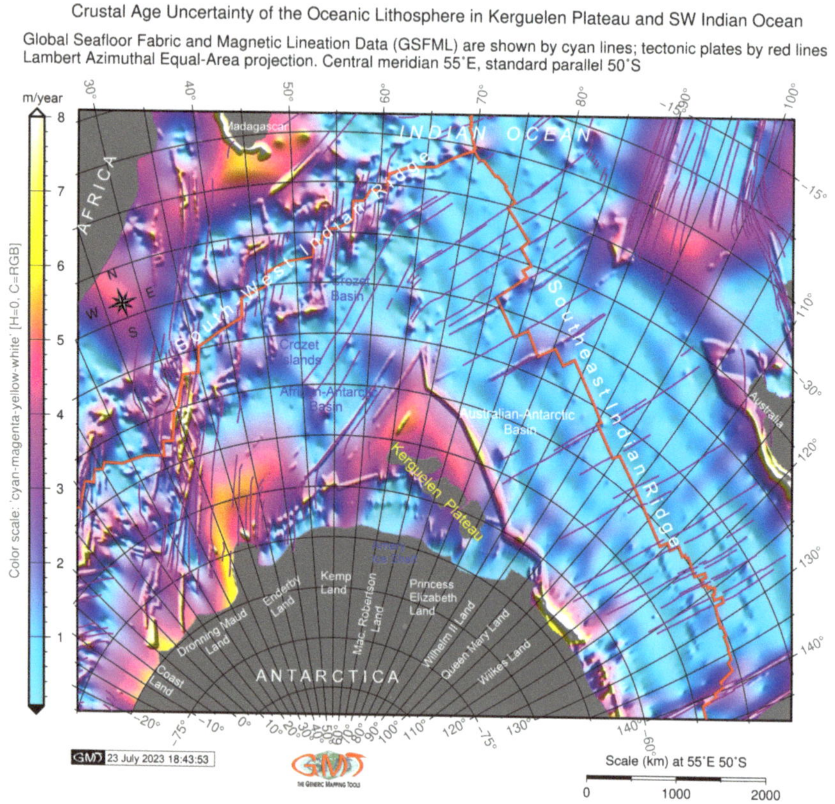

Figure 5. Age uncertainty in lithosphere crust (m.y) over the Kerguelen Plateau region, east Antarctic and south-west Indian Ocean. Cartography: GMT. Map source: author.

3.2.3. Mapping Sediment Thickness

A representation of the GlobSed 5-arc-minute total sediment thickness grid is extracted from the global raster grid on a target location of the Kerguelen Archipelago area and surrounding regions using 'grdcut' command, which selects the file. The amplitude of values is checked using the Geospatial Data Abstraction Library (GDAL) library embedded in GMT using the 'gdalinfo'. Based on the range of the values, the colour palette table is defined accordingly and presented using the 'psscale' module. This module is configured with parameters specifying the geographic location for each element on the map. The '-Dg' command is employed to define the position of the colour scale on the map using coordinates. Furthermore, grid annotations and graticules are added utilizing the 'psbasemap' GMT module. The remaining part of the script, complete with added comments for concise explanations, is outlined in Listing A6 within the Appendix A of this study. This particular listing was utilized to create the map featured in Figure 6.

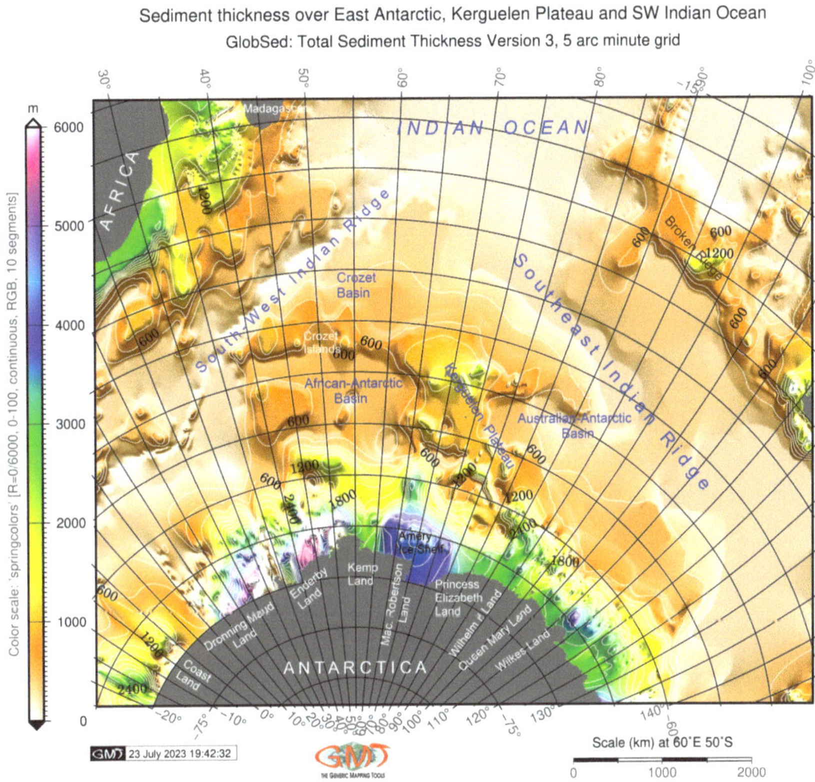

Figure 6. Sediment thickness (m) in East Antarctica, South-West Indian Ocean and the Kerguelen Plateau region. Cartography: GMT. Lambert azimuthal equal-area projection. Map source: author.

3.2.4. Geophysical Mapping

To emphasize features and subsurface characteristics that might not be apparent from geological data alone, maps of gravity anomalies were generated using gravity datasets projected onto the Lambert Equal-Area Azimuthal Projection for consistency with other maps. The complete scripts for creating these geophysical maps can be found in Appendix A, with Listing A7 detailing the process for free-air gravity anomalies and Listing A8 for vertical gravity gradient. The crucial lines of code are summarized below.

For free-air gravity anomalies, the 'img2grd' module was initially employed to convert the image from the .IMG format into GRD format. The projection was defined using the '-JU43/6.0i' flag within the 'grdimage' module. The extent of the region was specified using the '-R' command, and this information was subsequently passed to subsequent modules using the '-R -J' flags.

In the case of the vertical gravity gradient map, a satellite-derived grid sourced from altimeters CryoSat-2 and Jason-1/2 [115] was used. This dataset facilitated the identification of ridge propagation on the seafloor.

Due to GMT's ability to process each feature only once and utilize relevant modules ('psbasemap', 'grdcontour', 'psscale'), it boasts a high processing speed. This technical advantage empowers GMT to efficiently handle grids encompassing extensive geographic areas in a matter of seconds. With an understanding of the seafloor's geological and geophysical structure, GMT demonstrates rapidity and proficiency in mapping.

By adjusting projection parameters, the algorithm concentrates on key seafloor features within the desired scale and projection. Additionally, the GMT-based framework excels in determining the location, scope, and boundaries of mid-ocean ridges and significant basins.

Through cartographic generalization involving the gap in isolines, redundant bathymetric features can be eliminated. This process identifies pertinent and nonrepetitive features, underscores correlations between geophysical settings, gravity, and magnetic anomalies, and facilitates their comparison with geodetic anomalies. Consequently, GMT offers an effective approach to map and visualize seafloor structures.

3.2.5. Mapping Magnetic Anomalies

To map the magnetic anomalies, two different grids were used—the WDMAM and the EMAG2. The first was a subset from the global grid embedded in the GMT using the 'grdcut' module, and the second was obtained from the existing raster file of EMAG 2-min resolution grid. As a result, GMT offers flexibility in data processing through two approaches: utilizing pre-existing stored files and accessing embedded grids available within the system remotely. Subsequently, the images were rendered using the 'grdimage' module, accompanied by color palettes tailored to the data distribution range across the Kerguelen Plateau and the adjacent southwestern Indian Ocean. The 'makecpt' module was employed for this purpose. For a comprehensive view of the GMT scripts for both maps visualized with WDMAM and EMAG2, refer to Listings A9 and A10 in Appendix A.8.

3.2.6. Mapping Geoid Anomalies

The map of geoid inundations was created using the EGM96 dataset through the GMT script, with each module defining specific elements on the map. To achieve effective mapping, the scripts were executed from the console, tailoring the geoid plotting to target concepts such as projections, grids, colour palettes, raster extents, and annotations. The stepwise definition of these cartographic elements is evident on the geoid maps, as demonstrated in the script provided in Listing A11. This approach to data visualization ensures high levels of automation in data processing through GMT, surpassing conventional tools.

Significant features such as geoid undulations, seafloor isochrons, basement depth isolines, and asymmetries in crustal accretion are influenced by the asthenospheric flow from Kerguelen's mantle plumes to the spreading Southwest and Southeast Ridges. Mapping these data reveals insights into seafloor attributes and enables comparisons with the geological structure and geodynamics. Consequently, mapping multi-source topographic and geophysical datasets allows for the comparison of features using consistent projections. However, it is important to note that real-world seafloor features are seldom identical due to various factors impacting bathymetry, although they may exhibit strong correlations. To address this, GMT scripts can be adapted to different scales to map features in more detailed views. In order to assess the influence of geophysical and geological settings on seafloor structure using GMT's capabilities, numerous cartographic re-projections were generated. Each map represented selected seafloor features at varying levels of detail. The Lambert azimuthal equal-area projection was ultimately chosen as the optimal projection for its applicability across all maps, ensuring comparability and coherence.

4. Results and Discussion

Automatic matching of topographic and geophysical grids with high accuracy is essential for complex geologic–tectonic investigations. This paper demonstrated the use of scripting algorithms of GMT. These algorithms were applied for plotting eleven thematic maps covering the south-west segment of Indian Ocean and East Antarctic using bathymetric, geodynamic and high-resolution geophysical datasets on gravity (Figures 7 and 8). Cartographic scripts by GMT, as demonstrated in this research, provide visualized information on the geodynamic and geophysical setting of remotely located areas such as Kerguelen Plateau. During the cartographic process and workflow, the scripts enable us to save time through the increased speed of mapping due to the high level of automation.

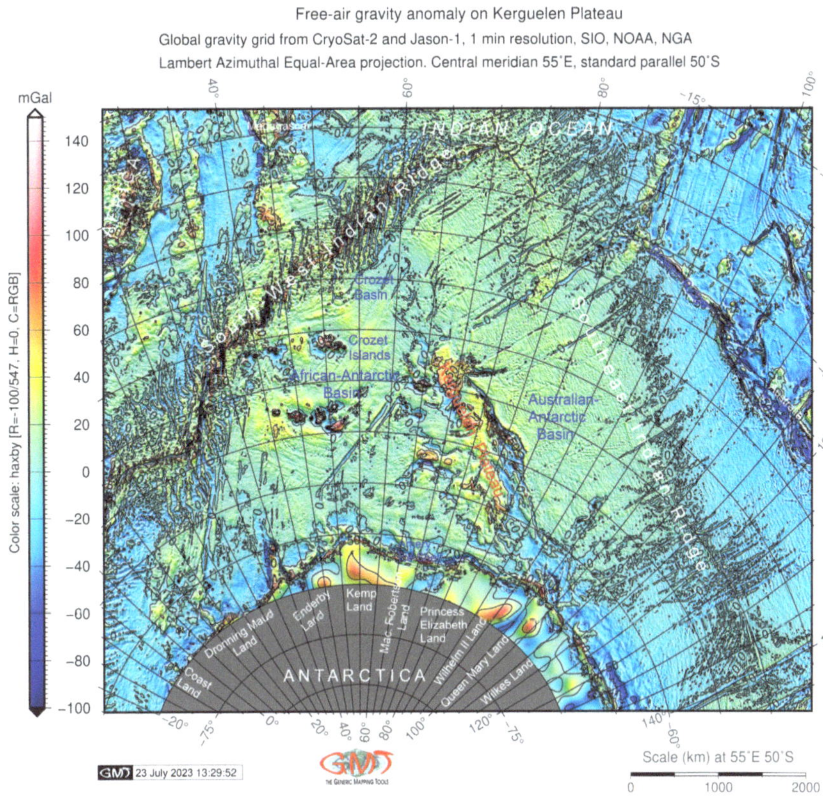

Figure 7. Marine gravity field over the Kerguelen Plateau region, south-west Indian Ocean. Mapping: GMT. Map source: author.

The programming concept of GMT enables us to better tune and adjust the layout of the cartographic plots in various scales and focus on a specific area for comparability of maps in a series. Such compatibility facilitates the evaluation of correlations among various geophysical and bathymetric features that have developed over the extensive history of the Kerguelen Archipelago. Furthermore, by utilizing geospatial analysis as a complementary technique, it becomes possible to compare and analyze the relationships within the geodynamic setting, such as age, spreading half rates of the ocean crust, asymmetry in crustal accretion on conjugate ridge flanks, and other variables. In this manner, GMT scripts offer an advanced cartographic method for visualizing datasets and extracting information through efficient data processing and modelling of geophysical properties of the seafloor.

These maps reveal the details of the structure of the seafloor in the Kerguelen Plateau. The results confirm that the dynamics of seafloor development, as reflected in the maps of oceanic crust age, asymmetries, and spreading rates of the south-west and southeast Indian ridges, are closely related to the major geophysical setting as depicted on the topographic, magnetic, and gravity grids. Additionally, the volcanic activity of the Kerguelen hotspot has a significant impact on the distribution of magnetic anomalies, which aligns with previous studies [116–118]. Overall, the results demonstrate that GMT scripting is a powerful and stable cartographic method that efficiently performs geophysical and bathymetric seafloor mapping. In the sections below, we discuss the obtained results on relevant maps, providing comments on the essential features and characteristics of the seafloor around the Kerguelen Plateau, south-west Indian Ocean, and East Antarctic.

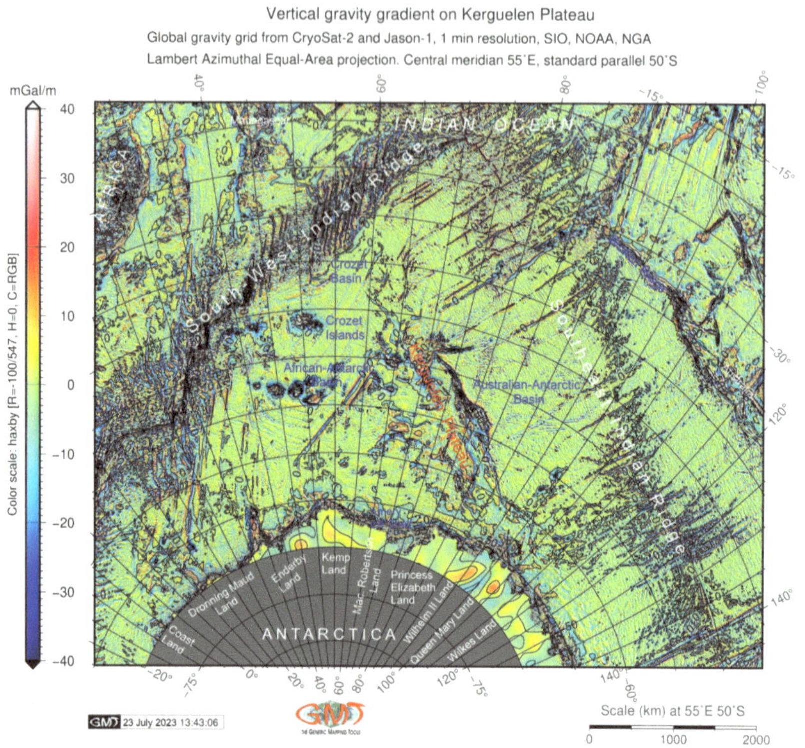

Figure 8. Vertical gravity gradient over Kerguelen Plateau. Mapping: GMT. Map source: author.

4.1. Ocean Floor Formation

The advanced methods of visualization by GMT constitute an important element of the map content through detailed plotting of the depicted objects, which enables us to indicate qualitative and quantitative geophysical specifications for reference and analysis. Thus, the general physical-geographic structure of the Kerguelen Plateau is visible in Figure 1, which shows a morphological orientation of the archipelago in the NW-SE direction and an extent surpassing 2000 km in length. As can be seen, the northern and southern parts of the plateau are asymmetric, where the less expressed southern part is older and lies in deeper water in the topographically downlifted areas. Age, spreading rates, and spreading symmetry of the ocean crust indicate the gradual evolution steps of the ocean floor formation in the southwest Indian Ocean and around the Kerguelen Plateau. The geodynamic setting of the oceanic crust (Figures 2–5) shows a strong relationship between the Kerguelen Plateau and the two mid-ocean ridges, which can be revealed from the analysis of the relief.

Relief is the main element of the seafloor since it reflects its geological structure and geodynamic history. Accordingly, the relief of the seafloor surface around the Kerguelen Plateau forms a continuously changing field of bathymetric heights. There are also sharp changes in altitude around the archipelago and mid-oceanic ridges. To depict the relief, GMT enables the modeling of isolines using the 'grdcontour' module and adjusted color gradients using methods of qualitative coloring of background and gradients according to the actual heights. At the same time, there are specific benefits of GMT techniques for mapping the hypsometric maps. Thus, the quantitative values of the relief make it possible to obtain absolute heights and elevations from the raster grid; the characteristics of the curvature and steepness of inclination can be obtained using the GMT module 'grdtrack' through cross-sectioning [119–121]. Moreover, GMT enables the modeling of the plasticity of the relief, that is, to depict a nonlinearity of the landform irregularities that form a visual image of the submarine terrain. This enables an analysis of the morphological conformity of the relief, which highlights major seafloor features, specific landforms, and their structure.

The age of the ocean crust was determined by interpolating the adjacent seafloor isochrons oriented towards the direction of seafloor spreading, as shown in Figure 2. This correspondence highlights the unique geophysical setting of the Kerguelen Plateau, underlain by the oceanic crust, which is strongly associated with its tectonic origin associated with volcanic hotspot and geologic history. The formation of the Southwest and the Southeast Indian Ridges is related to the uplift of the Kerguelen Plateau as a remnant of the Mesozoic oceanic basin existing after the separation of Gondwana. The comparison of the bathymetric map with geodynamic maps shows that seafloor heterogeneity around the Kerguelen Archipelago correlates with the seafloor spreading rates, where rougher basement is formed in the areas with the low half-spreading rate threshold (30–35 mm/year and lower). This correlation can be revealed by comparing the maps in Figures 1 and 4. Such heterogeneity in seafloor patterns varies significantly in various basins of the Indian Ocean, depending on the geodynamic setting and the geologic development of the oceanic crust.

The gridded map of age of the ocean crust around the Kerguelen Plateau (Figure 2) shows a correlation with the observed spreading half rates of the lithosphere (Figure 4). Originally formed as a single structure, Kerguelen was then split by the seafloor spreading in the south-west sector of the Indian Ocean which resulted in the formation of the two segments of the archipelago [122] which are visible on the maps. Furthermore, the difference in volcanic activity between the northern and central parts of the Kerguelen Plateau, underlying Heard Island, indicates that it is located on a hotspot, with various parts of the islands experiencing the effects of the mantle plume on different scales [123–125]. This unstable position has an impact on the geodynamic patterns over the Kerguelen Islands, leading to a higher uncertainty in the age of the oceanic crust (Figure 5) compared to the adjacent areas. Additionally, the structure of the oceanic crust beneath the Kerguelen Plateau is similar to that beneath aseismic ridges such as the Crozet Rise and the Madagascar Ridge, providing evidence that it originates from active volcanism associated with a hot spot [126].

4.2. Sediment Thickness

Mapping sediment thickness using the 5-arc-minute GlobSed grid relies on the approximated modelling, which highlights the sedimentation trends around the Kerguelen Archipelago, Figure 6.

The analytical map showing sediment thickness (Figure 6) displays patterns and key characteristics of sediment distribution over the seafloor. Using GMT-based techniques of mapping, it is easy to highlight the variations in sediment accumulation in various parts of the ocean using an adjusted colour table and the actual range in the data on sediment thickness. Hence, the level of cartographic details depends on the depth of the analysis with regard to the geologic formation of the seafloor and sediment accumulation. The objects on the map show the main regions of accumulated sediments, the structure and trends in distribution, and special features and properties compared to the closeness of coastal areas. Thus, the distribution of the sediment thickness correlates with the age of the underlying oceanic lithosphere and its latitude, which can be noted by the comparison of maps in Figures 2 and 6. Such correspondence is especially visible for higher values of the sediment thickness near the shorelines of the Antarctic, Amery Ice Shelf (3000–4000 m), Enderby Land (over 4000 m) and the Kerguelen Plateau (2000–3000 m).

A higher level of sediment thickness in these areas may also indicate earlier processes of subaerial erosion that occurred before subsidence and associated sedimentation. The rifting process that took place during the Late Paleogene resulted in changes in the Tertiary sediment facies of Kerguelen, which were influenced by the evolution of the Antarctic environment [127]. Sediments covering the Kerguelen Plateau include pillow lavas, tuffaceous sediments, and marine siltstones that were deposited since the late Miocene [128]. These sediments continue as a thick sequence of Cenozoic sediments (over 5000 m) within the Enderby Basin to the southwest of the Kerguelen Plateau (Figure 6).

The correlations observed between sediment thickness (Figure 6) and the age of the oceanic lithosphere (Figure 2) demonstrate the role of ocean floor formation in influencing the pattern of distribution and accumulation of sediments.

Moreover, the analytical map of sediment thickness reflects smaller features and details compared to the bathymetry of the southwest Indian Ocean. Hence, comparing the map of sediment thickness with the bathymetry enables us to detect associations, for example, high values of sediment thickness in the region of Dronning maud Land in the Antarctic, which can be associated with the effects from the processes of weathering and coastal erosion, factors of higher curvature in slopes and topographic variations in heights in the coastal areas. Other important factors increasing the accumulation of the sediments around the Kerguelen and the adjacent area include glacial processes and the turbidity of the ocean currents. Hence, intense circulation results in the accumulation of the large sediment fields with values over 3000 m.

These effects can be attributed to the Antarctic circumpolar currents that started around the Eocene-Oligocene periods and have continued until the present time. These currents constitute the strongest current system in the oceans, directed clockwise around the South Pole, and they significantly influence the adjacent sub-Antarctic regions, such as Kerguelen [129,130].

The general orientation of areas with maximal sedimentation is consistent with the sediment-filled troughs stretching in a NW-SE direction. These troughs are associated with the overall NW-SE orientation of Kerguelen and the axes of tectonic faults. The high values in sediment thickness around the Kerguelen Plateau, particularly contoured by the ridge isolines along its eastern margin, are associated with depositions resulting from bottom currents directed westwards.

4.3. *Free-Air Gravity Anomaly and Vertical Gravity Gradients*

The visualized marine gravity field over the Kerguelen Plateau region and the adjacent areas of the south-west Indian Ocean are shown in Figure 7. The comparison of the gravity roughness with the map of the half-spreading rates (Figure 4) and sediment thickness (Figure 6) shows the relationship between the speed of the spreading of mid-ocean ridges and roughness of the seafloor basement. Such phenomena are explained by the effects from the process of mid-ocean ridge formation. Other factors include the associated magma flows, spreading directions in Mesozoic and isochron orientations of the age of the oceanic crust, which affect current bathymetric and gravity patterns [131]. Furthermore, the gravity highs around the Kerguelen Plateau and Heard Island correspond to the maximal bathymetric elevations. These gravity highs indicate the presence of seamounts formed by Miocene basalts erupted during volcanic activity in the southern Indian Ocean. This volcanic activity contributed to the formation of a large igneous province [132].

The GMT-based geophysical maps enable us to determine the location and spatio-temporal structure of gravity phenomena that indicate on geological processes, their mutual relationships, and connections with topography. Such analysis supports the identification of trends and dynamics in seafloor development. It helps obtain quantitative characteristics from geophysical data and estimate both the highest and lowest values in gravity grids. In turn, zoning and classification of gravity variations helps to forecast changes in gravity anomalies over the seafloor of the Indian Ocean. Hence, the analysis of maps shows that the values of the free-air gravity over the Kerguelen are higher than in the surrounding areas and reach up to 80 mGal (Figure 8). In contrast, lower values are associated with the bathymetric depressions and have values of −40 to −60 mGal. Furthermore, the abyssal plain is characterized by the medium values of 0–20 mGal, Figure 8. This well illustrates the existing correlation of the free-air gravity anomalies with the distribution of topographic highs and depressions on the seafloor since they are strongly influenced by a gravitational effect of the distributed topographic masses that are caused by the differences in elevation.

Figure 8 displays the mapping outputs for the vertical gradient over the Kerguelen and the southwest Indian Ocean, showcasing the effects of different locations. The visualized map demonstrates a crucial property of gravitational systems, such as free-air gravity, which is not only subject to the effects of geographic location and the latitude of the selected measurement regions but also the altitude of the Earth's surface. This is because greater altitude implies a greater distance from the Earth's center, which in turn affects gravity values. Moreover, the vertical gravity gradient identifies variations in gravity with changes in topographic elevations, as depicted in Figure 8. The comparison of gravity datasets provides additional information on the distribution of major geological and seafloor structures, considering the variations of geophysical fields.

Furthremore, the comparison of the vertical gradient and free-air gravity map (Figures 7 and 8) with the topographic map (Figure 1) illustrates the effects of seafloor structure and the distribution of the oceanic bed on gravity, which shows that the highest values correspond to the Kerguelen Plateau and other rises, while lower values are generally associated with topographic depressions.

4.4. Magnetic Anomalies over Kerguelen Plateau

The anomaly of the magnetic intensity at an altitude of 5 km above mean sea level over the Kerguelen Plateau is shown in Figure 9.

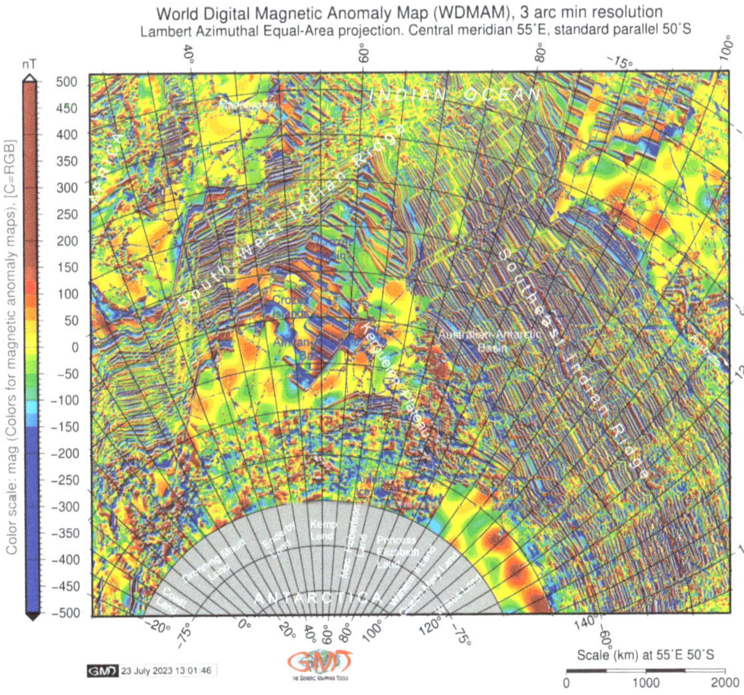

Figure 9. Patterns of the marine and terrestrial airborne magnetic anomaly over the Kerguelen Plateau on a three-minute resolution grid of WDMAM, south-west Indian Ocean. Map source: author.

Here, high heterogeneity in the geophysical data is related to the past volcanism over the Kerguelen Plateau, including the voluminous basaltic flooding originated from a deep hot spot as an asthenospheric source of mantle plume product. This resulted from the processes of slab dynamics and tectonic plate movements in the southwestern segment of the Indian Ocean. In this regard, combining the data from the WDMAM and EMAG2 (Figures 9 and 10) data on terrestrial gravity fields (Figures 7 and 8) for comparison with maps on oceanic crust development (Figures 2–5) presents an integrated GMT-based geophysical analysis.

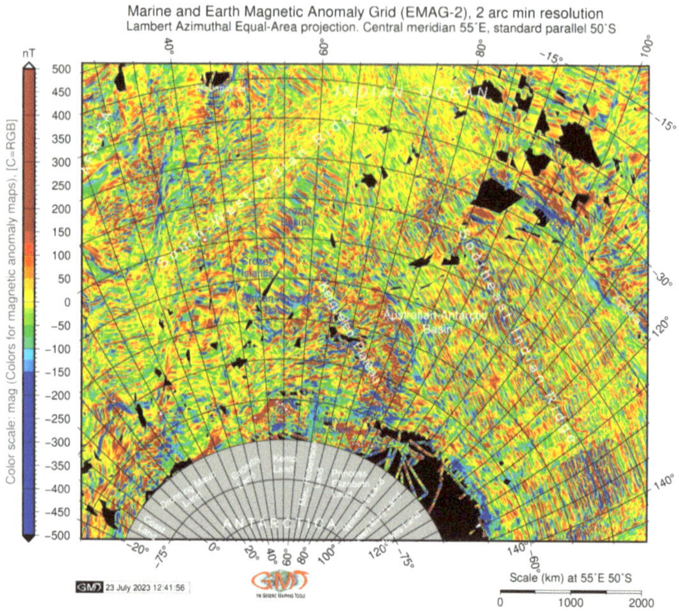

Figure 10. Marine and Earth airborne magnetic anomaly grid based on EMAG-2 over the Kerguelen Plateau region. Black areas signify "no data" in the original grid. Map source: author.

Crustal volume contributes to the decreased amplitude of the magnetic anomaly around the Kerguelen, as can be seen in Figure 10, with lower values of around −500 mGal. The analysis of the magnetic anomaly patterns in the SW Indian Ocean supports the hypothesis of the spreading seafloor with variations in the oceanic crustal block movements, as reported earlier [133]. This phenomenon is evident from the different magnetic patterns observed over the mid-ocean ridge. Moreover, the comparison of Figure 10 with Figure 5 reveals that seafloor age uncertainties for grid cells coincide with the marine magnetic anomaly identified around the Kerguelen Islands. This correlation is also observed for the conjugate ridge flanks (Figures 3 and 10).

The Earth Magnetic Anomaly Grid (EMAG2) offers the opportunity to assess magnetic-gravity field relationships as descriptors, going beyond the traditional analysis of gravity and magnetic anomalies. Magnetic anomalies arise from geological and topographic features that alter local magnetic fields, making it crucial to comprehend their correlation with geophysical phenomena and topography. For instance, scrutinizing local magnetic anomaly patterns in the southwest Indian Ocean reveals associations with oceanic crust formation, seafloor spreading, and subduction zones. Moreover, the age of oceanic crust and spreading rates, resulting from land accretion and extensive volcanism of the Kerguelen Plateau, are linked to the historical geological development of the region, as mentioned previously [134].

The EMAG-2 and WDMAM grids employed for plotting magnetic anomalies exhibit varying levels of grid detail, allowing insight into the subsurface structure of the seafloor around the Kerguelen Plateau and the composition of the Earth's crust in the southwest Indian Ocean. The magnetic fabric data correlate with hotspot activities and active volcanism, particularly prominent over the central and northern sectors of the archipelago [135–137]. Furthermore, intermediate crustal thickness values within the oceanic crust beneath the Kerguelen Plateau and large-amplitude magnetic anomalies across the archipelago point to the plateau's oceanic origin, attributed to plate volcanism resulting from tectonic plate activity, as previously reported [138,139]. Therefore, the distinctive magnetic patterns evident in Figure 10 maps correspond to heightened hotspot activity and the associated lava flows.

Deeper masses, asthenospheric upwelling, and mantle plume-driven convection are geodynamic processes that influence the magnetic properties of the Earth's surface. Moreover, through data analysis, a deeper understanding of the impact of geophysical settings on the distribution of positive and negative magnetic undulations emerges, with the former situated over the Kerguelen Islands and the latter in the eastern regions and southwest of Australia. This analysis enables the assessment of variations in geophysical grids through comparative map analysis. Thus, the cartographic depiction of geophysical and magnetic datasets offers advanced methods for extracting information about seafloor formation and interconnected geophysical processes.

4.5. Geoid Models

The geopotential model over the Kerguelen, based on the EGM96 dataset, is illustrated in Figure 11. The variations in the geoid across the Kerguelen Plateau highlight the ongoing isostatic compensation of the archipelago due to its low-density mantle. Consequently, the high anomalies in the geoid level above the Kerguelen Plateau can be attributed to the significant volcanic activity associated with the formation of ridges on the hot lithosphere. This volcanic activity is reflected in the exceptionally thick crust beneath the Kerguelen, resulting in geoid values exceeding 40 m (Figure 11), surpassing the normal thickness of the oceanic crust. These findings corroborate previous studies investigating the geoid in the Kerguelen Archipelago, which documented anomalous thickness in this region [140]. This isostatic compensation, linked to anomalously high geoid values, corresponds to the rugged elevated terrain in regions experiencing active tectonic uplift. These observations shed light on processes occurring in the upper mantle [141].

A comparison of the geoid map with topography (sf. Figures 1 and 11) implies an existing correlation between the continued geodynamic processes in the south-west Indian Ocean and the topographic structure of the Kerguelen Archipelago. Moreover, this proves a high positive correlation between the geoid height and deep structure of the seafloor topography, as also noted earlier for the regions of large plateaus and swells [142].

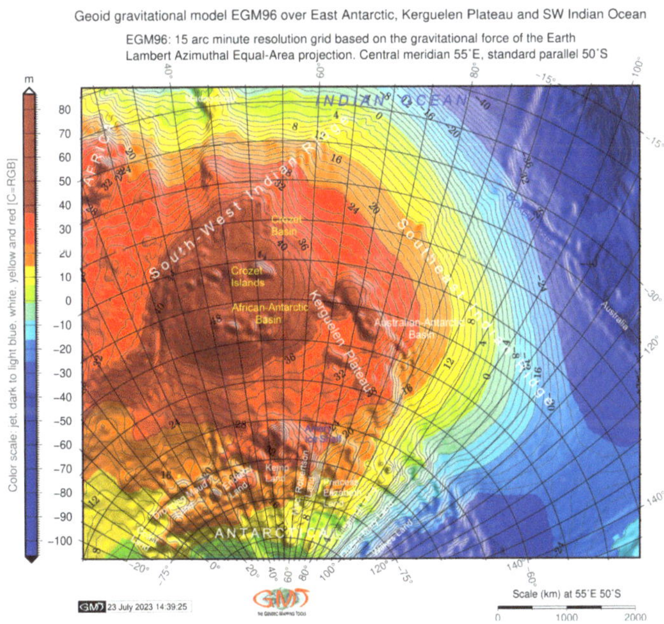

Figure 11. Geoid model based on EGM-96 over the Kerguelen Plateau region, east Antarctic and south-west Indian Ocean. Mapping: GMT. Map source: author.

5. Conclusions

As demonstrated in this study, the GMT-based mapping approach offers a wide range of cartographic functions for comprehensive spatio-temporal modelling and data visualization through an automated scripting approach. Utilizing GMT for cartographic tasks provides various modules and methods for representing different types of data, making it applicable in diverse fields of geomatics. The advantages of employing GMT in cartographic workflows are manifold. It enables highly automated plotting, facilitating rapid visualization of complex elements and features such as geographical, geological, oceanological, and geomorphological characteristics.

Furthermore, GMT supports multiple formats, encompassing both raster and vector data, and accommodating various classes and types of information. Leveraging the technical capabilities of GMT within cartographic workflows allows for common modelling and basic statistical analysis of spatial data, enhancing the understanding of their properties.

Analyzing a multitude of maps generated using GMT scripts reveals the consistency in depicted objects, facilitating their recognition and interpretation. The broad spectrum of GMT modules integrates scientific and technical methodologies for cartographic visualization and geospatial analysis. This unification aids in feature detection, recognition, and related research support.

Consequently, GMT-based mapping enables the amalgamation of maps for spatial analysis of intricate processes, objects, and phenomena, such as seafloor structures. This approach proves invaluable in addressing scientific and practical challenges within the realms of geophysics and geodynamics. The flexibility of GMT's syntax, the quality of its cartographic outputs, and its compatibility with various operating systems and computing devices all contribute to its effectiveness.

Given the success and applicability showcased in the executed GMT scripts–characterized by syntax flexibility, high-quality cartographic outputs, and compatibility across different platforms–it is foreseeable that this GMT-based cartographic method can be extended to study seafloor structures in other oceanic regions, considering varying geologic conditions and geodynamic evolutions.

Global surveillance of the seafloor through the use of altimeter satellites and gravity measurements has unveiled significant geophysical anomalies. Integrating data on magnetic field intensity, bathymetry, and deep seafloor geodynamics allows for an evaluation of the interrelations among these processes. This ongoing global surveillance continues to generate extensive high-resolution datasets. However, effectively analyzing these datasets in ever-higher spatial resolutions demands advanced tools for automated analysis. The toolset of GMT scripts has demonstrated its effectiveness in the realm of seafloor bathymetric and geophysical mapping. This approach facilitates the visualization and mapping of diverse seabed features across varying scales and resolutions, aiming to detect correlations between magnetic anomalies, geophysical patterns, and their connections to present bathymetry. Such visualization offers comprehensive coverage of seafloor features across the near-global scale.

Direct seafloor surveys for observations are resource-intensive, involving the use of complex and costly equipment, such as multi-beam echo-sounding systems, for data collection, generation, and storage. Yet, the need persists for efficient datasets that can be readily visualized and analyzed. Utilizing high-resolution geophysical datasets provided by NOAA and USGS, processed through the advanced scripting capabilities of GMT, establishes a cartographic processing pipeline for swift, automated, and accurate seafloor mapping. Notably, GMT's flexibility plays a pivotal role. A variety of GMT modules can be harnessed to process diverse geospatial data types, catering to different cartographic tasks. This flexibility allows for the adaptation and expansion of the proposed cartographic workflow, addressing larger-scale or smaller-scale mapping needs. The equilibrium between topographic gradients and geophysical grids vividly illustrates the links between seafloor patterns, the structure of the oceanic crust, and the processes within the lithospheric mantle.

From a cartographic perspective, this study underscores that the effective analysis of geologic-geophysical datasets within Earth sciences extends beyond utilizing isolated parameters (e.g., topographic maps) to encompass the selection of multiple datasets. The deep mantle processes, as reflected in geophysical data, intricately shape the seafloor formation. The methodology showcased in this study demonstrates how data from diverse sources (geophysical, topographic, geodetic, geodynamic) can be harnessed for comprehensive cartographic analysis using standardized workflows supported by scripts. This approach unveils additional insights into seafloor variability and the factors influencing ocean crust formation, strongly correlated with topographic patterns.

The cartographic approach exhibited here enables data assimilation and extension, not only across the Indian Ocean but also to other regions of the global ocean. For example, the Pacific Ocean boasts a rich tapestry of seafloor features, including vast abyssal plains, mid-ocean ridges, oceanic trenches, numerous seamounts, and continental shelves. These features present a fertile ground for investigating potential correlations between geophysical and magnetic anomalies and the heterogeneous seafloor patterns. In this context, the application of GMT for seafloor mapping in diverse oceanic regions serves as an ideal scenario for validating the cartographic scripting approach outlined in the methodological framework.

The series of maps presented, along with the comparison of the geophysical settings over the Kerguelen Plateau, underscores the superiority of script techniques over the GIS approach in terms of cartographic workflow automation. The compactness of GMT's syntax allows for code reusability with modifications. However, there are limitations to consider. GMT necessitates parameter tuning in advance when handling map elements and adjusting projection parameters, as it lacks the ability to preview maps before script execution, being a console-based program. Furthermore, GMT cannot remove redundant features once plotted, requiring the script to be run again for corrections. In contrast, the GIS approach permits real-time adjustments to map layouts, enabling the correction of colour palettes and bathymetric details on the fly. GMT's console-based nature mandates direct modifications to the script's code to address cartographic challenges.

Funding: This research received no external funding.

Institutional Review Board Statement: Not applicable.

Informed Consent Statement: Not applicable.

Data Availability Statement: Not applicable

Acknowledgments: The author sincerely thanks the two anonymous reviewers for critically reviewing this paper and for their constructive comments and useful suggestions, which have improved upon the initial version of this manuscript. The author is grateful to the Editorial Office of the *Geomatics* MDPI for support of this publication.

Conflicts of Interest: The authors declare no conflict of interest.

Abbreviations

The following abbreviations are used in this manuscript:

DEM	Digital Elevation Model
EMAG-2	Earth Magnetic Anomaly Grid 2
EML	ERDAS Macro Language
ESRI	Environmental Systems Research Institute
GDAL	Geospatial Data Abstraction Library
GRACE	Gravity Recovery and Climate Experiment
GMT	Generic Mapping Tools
GRASS GIS	Geographic Resources Analysis Support System
GUI	Graphical User Interface
NOAA	National Oceanic and Atmospheric Administration

SRTM	Shuttle Radar Topography Mission
USGS	United States Geological Survey
UTM	Universal Transverse Mercator
WGM	World Gravity Model
WDMAM	World Digital Magnetic Anomaly Map

Appendix A. GMT Scripts

Appendix A.1. GMT Script for Mapping Topography of the Kerguelen Region in Lambert Azimuthal Equal-Area Projection

Listing A1. Mapping topography of the Kerguelen region in Lambert Azimuthal Equal-Area projection.

```
# Subset study area by cutting a sub-region from a grid file
grdcut ETOPO1_Ice_g_gmt4.grd -R-40/150/-70/-20 -Gkgl_relief.nc
gdalinfo ss_relief.nc -stats
# Here the actual range is obtained from the 'gdalinfo'
# Minimum=-8239.000, Maximum=6392.000
# Color palette is used to access the master cpt tables and to translate them to fit the actual data range according to the z-values.
gmt makecpt -Cgray.cpt -V -T-8239/6392 > myocean.cpt
# Generate a file
ps=Bathymetry_Kgl.ps
gmt grdimage kgl_relief.nc -Cmyocean.cpt -R-25/-65/101/-10r -JA55/-50/7.5i -P -I+a15+ne0.75 -Xc -K > $ps
# Addiing shorelines by contouring the 2D gridded data sets
gmt grdcontour kgl_relief.nc -R -J -C1000 -Wthinnest,blue -O -K >> $ps
# Adding grid to create a basemap plot
gmt psbasemap -R -J \
    -Bpxg10f5a5 -Bpyg10f5a5 -Bsxg5 -Bsyg5 \
    -B+t"Topographic map of the Kerguelen Plateau" \
    -Lx15.0c/-1.3c+c318/-57+w2000k+l"Scale (km) at 60\232E 50\232S"+f \
    -UBL/-5p/-40p -O -K >> $ps
# Texts and various inscriptions on the maps are plotted using the 'pstext' module
gmt pstext -R -J -X0.0c -Y0.0c -N -O -K \
    -F+jTL+f9p,Helvetica,white+jLB >> $ps << EOF
66.0 -62.0 Enderby
66.5 -63.0 Basin
EOF
# Other text are added likewise
# Adding the legend is done using the 'psscale' module which explained the visualised conventional signs.
gmt psscale -Dg-27.0/-60+w15.4c/0.4c+v+ml -R -J -Cmyocean.cpt \
    -Bg1000f200a2000+l"Color scale: geo global bathymetry/topography relief [R=-8239/6392, H=0, C=RGB]" \
    -I0.2 -By+lm -O -K >> $ps
# Add GMT logo as depicted stamp of the software
gmt logo -Dx5.5/-2.2+o0.1i/0.1i+w2c -O -K >> $ps
# Add subtitle
gmt pstext -R0/10/0/15 -JX10/10 -X0.5c -Y13.0c -N -O \
    -F+f10p,Palatino-Roman,black+jLB >> $ps << EOF
4.0 6.1 ETOPO1 global terrain model, 1 arc min resolution grid
2.0 5.5 Lambert Azimuthal Equal-Area projection. Central meridian 55\232E, standard parallel 50\232S
EOF
# Convert to image file using GhostScript
gmt psconvert Bathymetry_Kgl.ps -A1.0c -E720 -Tj -Z
```

Appendix A.2. GMT Script for Mapping Age, Spreading Rates, Spreading Asymmetry and Age Uncertainty of the Ocean Crust

Listing A2. Mapping age of the oceanic lithosphere over Kerguelen Plateau and SW Indian Ocean.

```
gmt grdcut age.3.2.nc -R-40/150/-70/-10 -Gker_age.tif
gdalinfo ker_age.tif -stats
# Minimum=0.000, Maximum=16001.000, Mean=5895.761, StdDev=4042.817
# Color palette
# gmt makecpt -Cjet -T14/16000.000 > age.cpt
gmt makecpt -Cwysiwyg -T14/16000.000 > age.cpt
#gmt makecpt --help
# Generate a file
ps=Ker_age.ps
gmt grdimage ker_age.tif -Cage.cpt -R-25/-65/101/-10r -JA55/-50/7.5i -P -I
    +a15+ne0.75 -Xc -K > $ps
# add grid
gmt psbasemap -R -J \
    -Bpxg10f5a10 -Bpyg10f5a10 -Bsxg5 -Bsyg5 \
    -B+t"Age of Oceanic Lithosphere: Kerguelen Plateau and SW Indian Ocean
    " -O -K >> $ps
# Legend
gmt psscale -Dg-30/-58+w15.4c/0.4c+v+ml+e -R -J -Cage.cpt \
    -Bg1000f200a1000+l"Color scale: 'wysiwyg' [R=0/6000, H=0, C=RGB]" \
    -I0.2 -By+l"M.Y." -O -K >> $ps
# Scale, directional rose
gmt psbasemap -R -J \
    -Tdx0.8c/10.3c+w0.3i+f2+l+o0.15i \
    -Lx16.0c/-1.6c+c318/-57+w2000k+l"Scale (km) at 55\232E 50\232S"+f \
    -UBL/-5p/-40p -O -K >> $ps
# GMT logo
gmt logo -Dx5.5/-2.2+o0.1i/0.1i+w2c -O -K >> $ps
# Subtitle
gmt pstext -R0/10/0/15 -JX10/10 -X0.5c -Y13.0c -N -O \
    -F+f12p,0,black+jLB >> $ps << EOF
0.0 5.9 Lambert Azimuthal Equal-Area projection. Central meridian 55\232E,
     standard parallel 50\232S
0.0. 6.6 Age of the ocean crust, 2 arc min netCDF grid v.3., according to
    adjacent seafloor isochrons
EOF
# Convert to image file using GhostScript
gmt psconvert Ker_age.ps -A1.7c -E720 -Tj -Z
```

Appendix A.3. GMT Script for Mapping Asymmetries in Crustal Accretion on Conjugate Ridge Flanks: Kerguelen Plateau and SW Indian Ocean

Listing A3. Mapping asymmetries in crustal accretion on conjugate ridge flanks: Kerguelen Plateau and SW Indian Ocean.

```
exec bash
# Cut off raster image
gmt grdcut asym.3.2.nc -R-40/150/-70/-10 -Gker_asym.tif
gdalinfo ker_asym.tif -stats
# Minimum=14.000, Maximum=10000.000, Mean=5491.073, StdDev=1628.733
# Make color palette
gmt makecpt -Cturbo -T14/10000.000 > asym.cpt
# Generate a file
ps=Ker_asym.ps
gmt grdimage ker_asym.tif -Casym.cpt -R-25/-65/101/-10r -JA55/-50/7.5i -P
    -I+a15+ne0.75 -Xc -K > $ps
# Grid
gmt psbasemap -R -J \
    -Bpxg10f5a10 -Bpyg10f5a10 -Bsxg5 -Bsyg5 \
    -B+t"Asymmetries in crustal accretion (%) on conjugate ridge flanks:
    Kerguelen Plateau and SW Indian Ocean" -O -K >> $ps
# Legend
gmt psscale -Dg-30/-58+w15.4c/0.4c+v+ml+e -R -J -Casym.cpt \
    -Bg1000f200a1000+l"Color scale: 'no_green' [R=0/6000, H=0, C=RGB]" \
    -I0.2 -By+lm -O -K >> $ps
# Scale, directional rose
gmt psbasemap -R -J \
    -Tdx0.8c/10.3c+w0.3i+f2+l+o0.15i \
    -Lx16.0c/-1.6c+c318/-57+w2000k+l"Scale (km) at 55\232E 50\232S"+f \
    -UBL/-5p/-40p -O -K >> $ps
# GMT logo
gmt logo -Dx5.5/-2.2+o0.1i/0.1i+w2c -O -K >> $ps
# Subtitle
gmt pstext -R0/10/0/15 -JX10/10 -X0.5c -Y13.0c -N -O \
    -F+f12p,0,black+jLB >> $ps << EOF
0.0 5.9 Lambert Azimuthal Equal-Area projection. Central meridian 55\232E,
     standard parallel 50\232S
0.0. 6.6 Age, spreading rates and spreading asymmetry of the ocean crust,
    2 arc min netCDF grid v.3
EOF
# Convert to image file using GhostScript
gmt psconvert Ker_asym.ps -A1.7c -E720 -Tj -Z
```

Appendix A.4. GMT Script for Mapping Spreading Half Rates of the Oceanic Lithosphere in Kerguelen Plateau and SW Indian Ocean

Listing A4. Mapping spreading half rates of the oceanic lithosphere in Kerguelen Plateau and SW Indian Ocean.

```
# GMT set up
gmt set FORMAT_GEO_MAP=dddF \
    MAP_FRAME_PEN=dimgray \
    MAP_FRAME_WIDTH=0.1c \
    MAP_TITLE_OFFSET=0.5c \
    MAP_ANNOT_OFFSET=0.1c \
    MAP_TICK_PEN_PRIMARY=thinner,dimgray \
    MAP_GRID_PEN_PRIMARY=thin,white \
    MAP_GRID_PEN_SECONDARY=thinnest,white \
    FONT_TITLE=12p,0,black \
    FONT_ANNOT_PRIMARY=7p,0,dimgray \
    FONT_LABEL=7p,0,dimgray
#Overwrite defaults of GMT
gmtdefaults -D > .gmtdefaults
exec bash
# Cut off raster image
gmt grdcut rate.3.6.nc -R-40/150/-70/-10 -Gker_rate.tif
gdalinfo ker_rate.tif -stats
# Minimum=0.000, Maximum=15000.000, Mean=3175.093, StdDev=2125.065
# Make color palette
gmt makecpt -Cf-30-31-32.cpt -T0/15000/250 -N -Iz > age.cpt
# Generate a file
ps=Ker_rate.ps
gmt grdimage ker_rate.tif -Cage.cpt -R-25/-65/101/-10r -JA55/-50/7.5i -P -
    I+a15+ne0.75 -Xc -K > $ps
# Add grid
gmt psbasemap -R -J \
    -Bpxg10f5a10 -Bpyg10f5a10 -Bsxg5 -Bsyg5 \
    -B+t"Spreading Half Rates of the Oceanic Lithosphere in Kerguelen
    Plateau and SW Indian Ocean" -O -K >> $ps
gmt psxy -R -J ridge.gmt -Sf0.5c/0.15c+l+t -Wthin,yellow -Gpurple -O -K >>
    $ps
gmt psxy -R -J TP_Indian.txt -L -Wthickest,red -O -K >> $ps
gmt psxy -R -J TP_Australian.txt -L -Wthickest,red -O -K >> $ps
gmt psxy -R -J TP_African.txt -L -Wthickest,red -O -K >> $ps
# tectonic slab contours
gmt psxy -R -J GSFML_SF_FZ_KM.gmt -Wthick,cyan2 -O -K >> $ps
gmt psxy -R -J GSFML_SF_FZ_RM.gmt -Wthick,cyan2 -O -K >> $ps
# transform faults
gmt psxy -R -J transform.gmt -Sc0.05c -Ggreen -Wthick,deeppink1 -O -K >>
    $ps
# Add legend
gmt psscale -Dg-30/-58+w15.4c/0.4c+v+ml+e -R -J -Cage.cpt \
    -Bg1000f200a1000+l"Color scale: 'f-30-31-32.cpt' from Gnuplot [R
    =0/15000, H=0, C=RGB]" \
    -I0.2 -By+l"mm/yr." -O -K >> $ps
# Add scale, directional rose
gmt psbasemap -R -J \
    -Tdx0.8c/10.3c+w0.3i+f2+l+o0.15i \
    -Lx16.0c/-1.6c+c318/-57+w2000k+l"Scale (km) at 55\232E 50\232S"+f \
    -UBL/-5p/-40p -O -K >> $ps
# Add GMT logo
gmt logo -Dx5.5/-2.2+o0.1i/0.1i+w2c -O -K >> $ps
# Add subtitle
gmt pstext -R0/10/0/15 -JX10/10 -X0.5c -Y13.0c -N -O \
    -F+f12p,0,black+jLB >> $ps << EOF
0.0 5.9 Lambert Azimuthal Equal-Area projection. Central meridian 55\232E,
    standard parallel 50\232S
0.0. 6.6 Global Seafloor Fabric and Magnetic Lineation Data (GSFML) are
    shown by cyan lines
EOF
# Convert to image file using GhostScript
gmt psconvert Ker_rate.ps -A1.7c -E720 -Tj -Z
```

Appendix A.5. GMT Script for Mapping Crustal Age Uncertainty of the Oceanic Lithosphere in Kerguelen Plateau and SW Indian Ocean

Listing A5. Mapping crustal age uncertainty of the oceanic lithosphere in Kerguelen Plateau and SW Indian Ocean.

```
#!/bin/sh
# Age, spreading rates and spreading asymmetry of the ocean crust, 2 arc
    min netCDF grid v 3
# GMT modules: gmtset, gmtdefaults, grdcut, makecpt, grdimage, psscale,
    grdcontour, psbasemap, gmtlogo, psconvert
exec bash
# Cut off raster image
gmt grdcut ageerror.3.2.nc -R-40/150/-70/-10 -Gker_error.tif
gdalinfo ker_error.tif -stats
# Minimum=8.000, Maximum=1313.000, Mean=196.616, StdDev=156.744
# Color palette
gmt makecpt -Ccyan-magenta-yellow-white.cpt -T8/800.000 -N > age.cpt
# Generate a file
ps=Ker_error.ps
gmt grdimage ker_error.tif -Cage.cpt -R-25/-65/101/-10r -JA55/-50/7.5i -P
    -I+a15+ne0.75 -Xc -K > $ps
# Add grid
gmt psbasemap -R -J -Bpxg10f5a10 -Bpyg10f5a10 -Bsxg5 -Bsyg5 \
    -B+t"Crustal Age Uncertainty of the Oceanic Lithosphere in Kerguelen
    Plateau and SW Indian Ocean" -O -K >> $ps
# gmt psxy -R -J ridge.gmt -Sf0.5c/0.15c+l+t -Wthin,yellow -Gpurple -O -K
    >> $ps
gmt psxy -R -J TP_Indian.txt -L -Wthickest,red -O -K >> $ps
gmt psxy -R -J TP_Australian.txt -L -Wthickest,red -O -K >> $ps
gmt psxy -R -J TP_African.txt -L -Wthickest,red -O -K >> $ps
# tectonic slab contours
# gmt psxy -R -J GSFML_SF_FZ_KM.gmt -Wthick,slateblue3 -O -K >> $ps
gmt psxy -R -J GSFML_SF_FZ_KM.gmt -Wthick,darkmagenta -O -K >> $ps
gmt psxy -R -J GSFML_SF_FZ_RM.gmt -Wthick,darkmagenta -O -K >> $ps
# transform faults
gmt psxy -R -J transform.gmt -Sc0.05c -Ggreen -Wthick,deeppink1 -O -K >>
    $ps
# Legend
gmt psscale -Dg-30/-58+w15.4c/0.4c+v+ml+e -R -J -Cage.cpt \
    --FONT_LABEL=10p,0,dimgray --FONT_ANNOT_PRIMARY=10p,0,black \
    -Bg100f20a100+l"Color scale: 'jet' [R=0/15000, H=0, C=RGB]" \
    -I0.2 -By+l"m/yr." -O -K >> $ps
# Scale, directional rose
gmt psbasemap -R -J \
    --FONT=10p,0,black --MAP_TITLE_OFFSET=0.3c \
    -Tdx0.8c/10.3c+w0.3i+f2+l+o0.15i \
    -Lx16.0c/-1.6c+c318/-57+w2000k+l"Scale (km) at 55\232E 50\232S"+f \
    -UBL/-5p/-40p -O -K >> $ps
# GMT logo
gmt logo -Dx5.5/-2.2+o0.1i/0.1i+w2c -O -K >> $ps
# Add subtitle
gmt pstext -R0/10/0/15 -JX10/10 -X0.5c -Y13.0c -N -O \
    -F+f12p,0,black+jLB >> $ps << EOF
0.0 5.9 Lambert Azimuthal Equal-Area projection. Central meridian 55\232E,
     standard parallel 50\232S
0.0. 6.6 Global Seafloor Fabric and Magnetic Lineation Data (GSFML) are
    shown by cyan lines
EOF
# Convert to image file using GhostScript
gmt psconvert Ker_error.ps -A1.7c -E720 -Tj -Z
```

Appendix A.6. GMT Script for Sediment Thickness over the Kerguelen Plateau and SW Indian Ocean

Listing A6. Mapping sediment thickness over the Kerguelen Plateau and SW Indian Ocean.

```
exec bash
#gmt grdcut GlobSed-v2.nc -R-25/-65/101/-10r -Gker_sed.nc
gmt grdcut GlobSed-v2.nc -R-40/150/-70/-10 -Gker_sed.nc
gdalinfo ker_sed.nc -stats
# Minimum=0.000, Maximum=8116.000, Mean=530.333, StdDev=728.622
# Make color palette
gmt makecpt -Cno_green.cpt -V -T0/6000 > myocean.cpt
# gmt makecpt --help
# Generate a file
ps=SedThick_Kgl.ps
gmt grdimage ker_sed.nc -Cmyocean.cpt -R-25/-65/101/-10r -JA55/-50/7.5i -P -I+a15+ne0.75 -Xc -K > $ps
# Add shorelines
gmt grdcontour ker_sed.nc -R -J -C200 -A200+f10p,25,black -Wthinner,dimgray -O -K >> $ps
# Add grid
gmt psbasemap -R -J \
    -Bpxg10f5a10 -Bpyg10f5a10 -Bsxg5 -Bsyg5 \
    -B+t"Sediment thickness over East Antarctic, Kerguelen Plateau and SW Indian Ocean" \
    -Lx15.0c/-1.5c+c318/-57+w2000k+l"Scale (km) at 60\232E 50\232S"+f \
    -UBL/-5p/-40p -O -K >> $ps
# Texts
# Add legend
gmt psscale -Dg-33/-59+w15.4c/0.4c+v+ml+e -R -J -Cmyocean.cpt \
    -Bg1000f50a1000+l"Color scale: 'no_green' [R=0/6000, H=0, C=RGB]" \
    -I0.2 -By+lm -O -K >> $ps
# Add GMT logo
gmt logo -Dx5.5/-2.2+o0.1i/0.1i+w2c -O -K >> $ps
# Add subtitle
gmt pstext -R0/10/0/15 -JX10/10 -X0.5c -Y13.0c -N -O \
    -F+f12p,0,black+jLB >> $ps << EOF
3.0 6.1 GlobSed: Total Sediment Thickness Version 3, 5 arc minute grid
EOF
# Convert to image file using GhostScript
gmt psconvert SedThick_Kgl.ps -A1.5c -E720 -Tj -Z
```

Appendix A.7. GMT Scripts for Geophysical Mapping over the Kerguelen Plateau

Listing A7. Mapping free-air gravity anomalies over Kerguelen Plateau and SW Indian Ocean.

```
exec bash
gmt img2grd grav_27.1.img -R40/110/-70/-20 -Ggrav_Ker.grd  T1  I1  E  S0.1 -V
gdalinfo grav_Ker.grd -stats
gmt makecpt -Chaxby -V -T-100/150 > myocean.cpt
ps=Gravity_Kgl.ps
gmt grdimage grav_Ker.grd -Cmyocean.cpt -R40/110/-70/-20 -JU43/6.0i -P -I+a15+ne0.75 -Xc -K > $ps
gmt grdcontour grav_Ker.grd -R -J -C50 -A50 -Wthinner -O -K >> $ps
gmt psbasemap -R -J \
    -Bpxg10f5a10 -Bpyg10f5a5 -Bsxg5 -Bsyg5 \
    -B+t"Free-air gravity anomaly on Kerguelen Plateau" \
    -Lx7.5c/-1.3c+c318/-57+w2000k+l"UTM projection, Zone 43. Scale (km)"+f \
    -UBL/10p/-40p -O -K >> $ps
gmt psscale -Dg41/-12.5+w15.4c/0.4c+ml+h+e -R -J -Cmyocean.cpt \
    -Bg20f2a20+l"Color scale: haxby [R=-100/547, H=0, C=RGB]" \
    -I0.2 -By+lm -O -K >> $ps
gmt logo -Dx7.5/-2.2+o0.1i/0.1i+w2c -O -K >> $ps
gmt pstext -R0/10/0/15 -JX10/10 -X0.5c -Y13.0c -N -O \
    -F+f10p,0,black+jLB >> $ps << EOF
1.0 1.3 Global gravity grid from CryoSat-2 and Jason-1, 1 min resolution, SIO, NOAA, NCA
EOF
gmt psconvert Gravity_Kgl.ps -A1.0c -E720 -Tj -Z
```

Listing A8. Mapping vertical gravity gradient over Kerguelen Plateau and SW Indian Ocean.

```
exec bash
gmt img2grd curv_27.1.img -R40/110/-70/-20 -Ggravvert_Ker.grd -T1 -I1 -E -S0.1 -V
gdalinfo gravvert_Ker.grd -stats
gmt makecpt -Cmag.cpt -T-40/40 > colors.cpt
ps=Gravity_Kgl_vert.ps
gmt grdimage gravvert_Ker.grd -Ccolors.cpt -R40/110/-70/-20 -JU43/6.0i -P -I+a15+ne0.75 -Xc -K > $ps
gmt grdcontour gravvert_Ker.grd -R -J -C100 -A100 -Wthinner -O -K >> $ps
gmt psbasemap -R -J \
    -Bpxg10f5a10 -Bpyg10f5a5 -Bsxg5 -Bsyg5 \
    -B+t"Vertical gravity gradient over Kerguelen Plateau" \
    -Lx7.5c/-1.3c+c318/-57+w2000k+l"UTM projection, Zone 43. Scale (km)"+f \
    -UBL/10p/-40p -O -K >> $ps
gmt psscale -Dg41/-12.5+w15.4c/0.4c+ml+h+e -R -J -Ccolors.cpt \
    -Bg20f2a20+l"Color scale: mag [R=-40/40, H=0, C=RGB]" \
    -I0.2 -By+l"mGal/m" -O -K >> $ps
gmt logo -Dx7.5/-2.2+o0.1i/0.1i+w2c -O -K >> $ps
# Add subtitle
gmt pstext -R0/10/0/15 -JX10/10 -X0.5c -Y13.0c -N -O \
    -F+f10p,0,black+jLB >> $ps << EOF
1.0 1.3 Global gravity grid derived from satellite altimetry (CryoSat-2 and Jason-1: NOAA, NGA)
EOF
gmt psconvert Gravity_Kgl_vert.ps -A1.0c -E720 -Tj -Z
```

Appendix A.8. GMT Scripts for Mapping Magnetic Anomalies over the Kerguelen Plateau

Listing A9. Mapping magnetic anomalies by WDMAM over Kerguelen Plateau.

```
exec bash
gmt grdcut @earth_wdmam_03m -R50/100/-70/-30 -Gker_mag.nc
gdalinfo ker_mag.nc -stats
gmt makecpt -Cmag.cpt -V -T-500/500 > myocean.cpt
ps=Magnet_Kgl.ps
gmt grdimage ker_mag.nc -Cmyocean.cpt -R50/100/-70/-30 -JM6.5i -P -I+a15+ne0.75 -Xc -K > $ps
gmt grdcontour ker_mag.nc -R -J -C200 -A1+f10p,25,black -Wthinner,dimgray -O -K >> $ps
gmt psbasemap -R -J \
    -Bpxg10f5a5 -Bpyg10f5a5 -Bsxg5 -Bsyg5 \
    -B+t"Marine and Earth airborne Magnetic Anomaly Grid on Kerguelen Plateau" \
    -Lx14.0c/-3.0c+c318/-57+w1000k+l"Mercator projection. Scale (km)"+f \
    -UBL/-5p/-80p -O -K >> $ps
gmt psscale -Dg50/-71.5+w16.0c/0.4c+h+ml+e -R -J -Cmyocean.cpt \
    -Bg100f10a50+l"Color scale: mag (Colors for magnetic anomaly maps), [C=RGB]" \
    -I0.2 -By+l"nT" -O -K >> $ps
gmt logo -Dx6.5/-3.8+o0.1i/0.1i+w2c -O -K >> $ps
gmt pstext -R0/10/0/15 -JX10/10 -X0.5c -Y13.0c -N -O \
    -F+f11p,0,black+jLB >> $ps << EOF
1.2 15.8 WDMAM (World Digital Magnetic Anomaly Map), 3 arc min resolution grid
EOF
gmt psconvert Magnet_Kgl.ps -A1.0c -E720 -Tj -Z
```

Listing A10. Mapping magnetic anomalies by EMAG2 over Kerguelen Plateau.

```
exec bash
gmt grdcut EMAG2_V2.grd -R50/100/-70/-30 -Gker_mag.nc
gdalinfo ker_mag.nc -stats
gmt makecpt -Cmag.cpt -V -T-500/500 > myocean.cpt
ps=Magnet_Kgl.ps
gmt grdimage ker_mag.nc -Cmyocean.cpt -R50/100/-70/-30 -JM6.5i -P -I+a15+
    ne0.75 -Xc -K > $ps
gmt grdcontour ker_mag.nc -R -J -C200 -A1+f10p,25,black -Wthinner,dimgray
    -O -K >> $ps
gmt psbasemap -R -J \
    -Bpxg10f5a5 -Bpyg10f5a5 -Bsxg5 -Bsyg5 \
    -B+t"Detailed marine and Earth EMAG-2 airborne Magnetic Anomaly Grid
    on Kerguelen Plateau" \
    -Lx14.0c/-3.0c+c318/-57+w1000k+l"Mercator projection. Scale (km)"+f \
    -UBL/-5p/-80p -O -K >> $ps
gmt psscale -Dg50/-71.5+w16.0c/0.4c+h+ml+e -R -J -Cmyocean.cpt \
    -Bg100f10a50+l"Color scale: mag (Colors for magnetic anomaly maps), [C
    =RGB]" \
    -I0.2 -By+l"nT" -O -K >> $ps
gmt logo -Dx6.5/-3.8+o0.1i/0.1i+w2c -O -K >> $ps
gmt pstext -R0/10/0/15 -JX10/10 -X0.5c -Y13.0c -N -O \
    -F+f11p,0,black+jLB >> $ps << EOF
1.2 15.8 EMAG-2 (Earth Magnetic Anomaly Grid, 2 arc min resolution)
EOF
gmt psconvert Magnet_Kgl.ps -A1.0c -E720 -Tj -Z
```

Appendix A.9. GMT Scripts for Mapping Geoid Anomalies over the Kerguelen Plateau

Listing A11. Mapping geoid anomalies in Kerguelen Plateau by EGM96.

```
exec bash
gdalinfo geoid.egm96.grd -stats
gmt grd2cpt geoid.egm96.grd -Cjet > geoid.cpt
ps=Geoid_Ker.ps
gmt grdimage geoid.egm96.grd -I+a45+nt1 -R40/110/-70/-20 -JQ7.5i -Cgeoid.
    cpt -P -K > $ps
gmt grdcontour geoid.egm96.grd -R -J -C2 -A4+f10p,25,black -Wthinner,
    dimgray -O -K >> $ps
gmt psbasemap -R -J \
    -Bpxg10f5a10 -Bpyg10f5a10 -Bsxg5 -Bsyg5 \
    -B+t"Geoid gravitational model EGM96 over East Antarctic, Kerguelen
    Plateau and SW Indian Ocean" \
    -Lx16.0c/-1.5c+c318/-57+w1000k+l"Cylindrical equidistant projection.
    Scale (km)"+f \
    -UBL/-5p/-40p -O -K >> $ps
gmt psscale -Dg33/-70+w13.4c/0.4c+v+ml+e -R -J -Cgeoid.cpt \
    -Bg10f1a10+l"Color scale: jet [C=RGB]" \
    -I0.2 -By+lm -O -K >> $ps
gmt logo -Dx7.0/-2.2+o0.1i/0.1i+w2c -O -K >> $ps
gmt pstext -R0/10/0/15 -JX10/10 -X0.5c -Y12.5c -N -O \
    -F+f12p,0,black+jLB >> $ps << EOF
1.0 3.0 EGM96: 15 arc minute resolution grid based on the gravitational
    force of the Earth
EOF
gmt psconvert Geoid_Ker.ps -A1.6c -E720 -Tj -Z
```

References

1. Papp-Váry, Á. The Science of Cartography. In *Cartography Past, Present and Future*; Rhind, D., Taylor, D., Eds.; International Cartographic Association, Pergamon: Amsterdam, The Netherlands, 1989; pp. 103–109. [CrossRef]
2. Robinson, A.H. Cartography as an art. In *Cartography Past, Present and Future*; Rhind, D., Taylor, D., Eds.; International Cartographic Association, Pergamon: Amsterdam, The Netherlands, 1989; pp. 91–102. [CrossRef]
3. Goodchild, M.F. Geographic Information Systems and Cartography. In *International Encyclopedia of Human Geography*, 2nd ed.; Kobayashi, A., Ed.; Elsevier: Oxford, UK, 2009; pp. 1–6. [CrossRef]
4. Goodchild, M. GIS and Cartography. In *International Encyclopedia of Human Geography*; Kitchin, R., Thrift, N., Eds.; Elsevier: Oxford, UK, 2009; pp. 500–505. [CrossRef]

5. Slocum, T.A.; Egbert, S.L. Cartographic Data Display. In *Geographic Information Systems*; Taylor, F., Ed.; Modern Cartography Series; Academic Press: Cambridge, MA, USA, 1991; Chapter 9, Volume 1, pp. 167–199. [CrossRef]
6. Taylor, D.F. Perspectives on Visualization and Modern Cartography. In *Visualization in Modern Cartography*; Maceachren, A.M., Taylor, D.F., Eds.; Modern Cartography Series; Academic Press: Cambridge, MA, USA, 1994; Chapter 17, Volume 2, pp. 333–341. [CrossRef]
7. Toms, S.; Parker, B.; Tucker, D.C.; Rubalcava, R. *Python for ArcGIS Pro: Automate Cartography and Data Analysis Using ArcPy, ArcGIS API for Python, Notebooks, and Pandas*; Packt Publishing: Birmingham, UK, 2022.
8. Dobesova, Z.; Dobes, P. Comparison of visual languages in Geographic Information Systems. In Proceedings of the 2012 IEEE Symposium on Visual Languages and Human-Centric Computing (VL/HCC), Innsbruck, Austria, 30 September–4 October 2012; pp. 245–246. [CrossRef]
9. Omran, A.; Dietrich, S.; Abouelmagd, A.; Michael, M. New ArcGIS tools developed for stream network extraction and basin delineations using Python and java script. *Comput. Geosci.* **2016**, *94*, 140–149. [CrossRef]
10. Naghibi, S.A.; Hashemi, H.; Pradhan, B. APG: A novel python-based ArcGIS toolbox to generate absence-datasets for geospatial studies. *Geosci. Front.* **2021**, *12*, 101232. [CrossRef]
11. Roberts, J.; Mwangi, R.; Mukabi, F.; Njui, J.; Nzioka, K.; Ndambiri, J.; Bispo, P.; Espirito-Santo, F.; Gou, Y.; Johnson, S.; et al. Pyeo: A Python package for near-real-time forest cover change detection from Earth observation using machine learning. *Comput. Geosci.* **2022**, *167*, 105192. [CrossRef]
12. Lemenkova, P.; Debeir, O. Satellite Image Processing by Python and R Using Landsat 9 OLI/TIRS and SRTM DEM Data on Côte d'Ivoire, West Africa. *J. Imaging* **2022**, *8*, 317. [CrossRef] [PubMed]
13. Vos, K.; Splinter, K.D.; Harley, M.D.; Simmons, J.A.; Turner, I.L. CoastSat: A Google Earth Engine-enabled Python toolkit to extract shorelines from publicly available satellite imagery. *Environ. Model. Softw.* **2019**, *122*, 104528. [CrossRef]
14. Lemenkova, P.; Debeir, O. R Libraries for Remote Sensing Data Classification by K-Means Clustering and NDVI Computation in Congo River Basin, DRC. *Appl. Sci.* **2022**, *12*, 12554. [CrossRef]
15. Dobesova, Z. Programming language Python for data processing. In Proceedings of the 2011 International Conference on Electrical and Control Engineering, Yichang, China, 16–18 September 2011; pp. 4866–4869. [CrossRef]
16. Lemenkova, P. Cartographic scripts for seismic and geophysical mapping of Ecuador. *Geografie* **2022**, *127*, 195–218. [CrossRef]
17. Aber, S.E.W.; Aber, J.W. Basic Map Concepts—The Science of Cartography. In *Map Librarianship*; Aber, S.E.W., Aber, J.W., Eds.; Chandos Information Professional Series; Chandos Publishing: Oxford, UK, 2017; Chapter 3, pp. 53–70. [CrossRef]
18. Bascou, J.; Delpech, G.; Vauchez, A.; Moine, B.N.; Cottin, J.Y.; Barruol, G. An integrated study of microstructural, geochemical, and seismic properties of the lithospheric mantle above the Kerguelen plume (Indian Ocean). *Geochem. Geophys. Geosyst.* **2008**, *9*, 1–26. [CrossRef]
19. Eshagh, M. Satellite gravimetry and isostasy. In *Satellite Gravimetry and the Solid Earth*; Eshagh, M., Ed.; Elsevier: Amsterdam, The Netherlands, 2021; Chapter 7, pp. 313–373. [CrossRef]
20. Kapawar, M.; Mamilla, V.; Sangode, S. Anisotropy of magnetic susceptibility study to locate the feeder zone and lava flow directions of the Rajmahal Traps (India): Implications to Kerguelen mantle plume interaction with Indian Plate. *Phys. Earth Planet. Inter.* **2021**, *313*, 106692. [CrossRef]
21. Peng, W.; Yang, T.; Shi, Y.; Bian, W.; Ma, Y.; Wang, S.; Jiao, X.; Ma, J.; Kang, Y.; Zhang, S.; et al. Role of the Kerguelen mantle plume in breakup of eastern Gondwana: Evidence from early cretaceous volcanic rocks in the eastern Tethyan Himalaya. *Palaeogeogr. Palaeoclimatol. Palaeoecol.* **2022**, *588*, 110823. [CrossRef]
22. Zhang, Z.; Li, G.; He, X.; Zhang, L.; Dong, S.; Qing, C.; Liang, W.; Han, S. The evolution of Kerguelen mantle plume and breakup of eastern Gondwana: New insights from multistage Cretaceous magmatism in the Tethyan Himalaya. *Gondwana Res.* **2023**, *119*, 68–85. [CrossRef]
23. Lemenkova, P. Variations in the bathymetry and bottom morphology of the Izu-Bonin Trench modelled by GMT. *Bull. Geography. Phys. Geogr. Ser.* **2020**, *18*, 41–60. [CrossRef]
24. Tesauro, M.; Kaban, M.K.; Cloetingh, S.A. Global model for the lithospheric strength and effective elastic thickness. *Tectonophysics* **2013**, *602*, 78–86. [CrossRef]
25. Battista, T.; O'Brien, K. Spatially Prioritizing Seafloor Mapping for Coastal and Marine Planning. *Coast. Manag.* **2015**, *43*, 35–51. [CrossRef]
26. Lemenkova, P. The visualization of geophysical and geomorphologic data from the area of Weddell Sea by the Generic Mapping Tools. *Stud. Quat.* **2021**, *38*, 19–32. [CrossRef]
27. Lemenkova, P. GEBCO Gridded Bathymetric Datasets for Mapping Japan Trench Geomorphology by Means of GMT Scripting Toolset. *Geod. Cartogr.* **2020**, *46*, 98–112. [CrossRef]
28. Schlich, R.; Wise, S.W. The Geologic and Tectonic Evolution of the Kerguelen Plateau: An Introduction to the Scientific Results of Leg 120. *Proc. Ocean. Drill. Program Sci. Results* **1992**, *120*, 4.
29. Veevers, J.J.; Li, Z.X. Review of seafloor spreading around Australia. II. Marine magnetic anomaly modelling. *Aust. J. Earth Sci.* **1991**, *38*, 391–408. [CrossRef]

30. Negi, S.S.; Kumar, A.; Ningthoujam, L.S.; Pandey, D.K. Mapping the mantle transition zone beneath the Indian Ocean geoid low from Ps receiver functions. *Tectonophysics* **2022**, *831*, 229330. [CrossRef]
31. Bénard, F.; Callot, J.P.; Vially, R.; Schmitz, J.; Roest, W.; Patriat, M.; Loubrieu, B. The Kerguelen plateau: Records from a long-living/composite microcontinent. *Mar. Pet. Geol.* **2010**, *27*, 633–649. [CrossRef]
32. Ramana, M.V.; Ramprasad, T.; Desa, M. Seafloor spreading magnetic anomalies in the Enderby Basin, East Antarctica. *Earth Planet. Sci. Lett.* **2001**, *191*, 241–255. [CrossRef]
33. Davis, J.K.; Lawver, L.A.; Norton, I.O.; Gahagan, L.M. New Somali Basin magnetic anomalies and a plate model for the early Indian Ocean. *Gondwana Res.* **2016**, *34*, 16–28. [CrossRef]
34. Frey, F.; Coffin, M.; Wallace, P.; Weis, D.; Zhao, X.; Wise, S.; Wähnert, V.; Teagle, D.; Saccocia, P.; Reusch, D.; et al. Origin and evolution of a submarine large igneous province: The Kerguelen Plateau and Broken Ridge, southern Indian Ocean. *Earth Planet. Sci. Lett.* **2000**, *176*, 73–89. [CrossRef]
35. Sreejith, K.M.; Krishna, K.S. Magma production rate along the Ninetyeast Ridge and its relationship to Indian plate motion and Kerguelen hot spot activity. *Geophys. Res. Lett.* **2015**, *42*, 1105–1112. [CrossRef]
36. Lemenkova, P. Topography of the Aleutian Trench south-east off Bowers Ridge, Bering Sea, in the context of the geological development of North Pacific Ocean. *Baltica* **2021**, *34*, 27–46. [CrossRef]
37. Falvey, D.A. Kerguelen Plateau Research Cruise. *Eos Trans. Am. Geophys. Union* **1985**, *66*, 641. [CrossRef]
38. Watson, S.; Whittaker, J.; Halpin, J.; Williams, S.; Milan, L.; Daczko, N.; Wyman, D. Tectonic drivers and the influence of the Kerguelen plume on seafloor spreading during formation of the early Indian Ocean. *Gondwana Res.* **2016**, *35*, 97–114. [CrossRef]
39. Lee, Y. Cartographic Data Capture and Storage. In *Geographic Information Systems*; Taylor, F., Ed.; Modern Cartography Series; Academic Press: Cambridge, MA, USA, 1991; Chapter 2, Volume 1, pp. 21–38. [CrossRef]
40. Mori, M.; Corney, S.P.; Melbourne-Thomas, J.; Welsford, D.C.; Klocker, A.; Ziegler, P.E. Using satellite altimetry to inform hypotheses of transport of early life stage of Patagonian toothfish on the Kerguelen Plateau. *Ecol. Model.* **2016**, *340*, 45–56. [CrossRef]
41. Cotté, C.; Ariza, A.; Berne, A.; Habasque, J.; Lebourges-Dhaussy, A.; Roudaut, G.; Espinasse, B.; Hunt, B.; Pakhomov, E.; Henschke, N.; et al. Macrozooplankton and micronekton diversity and associated carbon vertical patterns and fluxes under distinct productive conditions around the Kerguelen Islands. *J. Mar. Syst.* **2022**, *226*, 103650. [CrossRef]
42. Maraldi, C.; Lyard, F.; Testut, L.; Coleman, R. Energetics of internal tides around the Kerguelen Plateau from modeling and altimetry. *J. Geophys. Res. Ocean.* **2011**, *116*, 1–10. [CrossRef]
43. Verfaillie, D.; Favier, V.; Dumont, M.; Jomelli, V.; Gilbert, A.; Brunstein, D.; Gallée, H.; Rinterknecht, V.; Menegoz, M.; Frenot, Y. Recent glacier decline in the Kerguelen Islands (49°S, 69°E) derived from modeling, field observations, and satellite data. *J. Geophys. Res. Earth Surf.* **2015**, *120*, 637–654. [CrossRef]
44. Yamamoto, K.; Fukuda, Y.; Doi, K.; Motoyama, H. Interpretation of the GRACE-derived mass trend in Enderby Land, Antarctica. *Polar Sci.* **2008**, *2*, 267–276. [CrossRef]
45. Mathieu, L.; Byrne, P.; Guillaume, D.; van Wyk de Vries, B.; Moine, B. The field and remote sensing analysis of the Kerguelen Archipelago structure, Indian Ocean. *J. Volcanol. Geotherm. Res.* **2011**, *199*, 206–215. [CrossRef]
46. Cowley, S.; Mann, P.; Coffin, M.; Shipley, T.H. Oligocene to Recent tectonic history of the Central Solomon intra-arc basin as determined from marine seismic reflection data and compilation of onland geology. *Tectonophysics* **2004**, *389*, 267–307. [CrossRef]
47. Lemenkova, P. Dataset compilation by GRASS GIS for thematic mapping of Antarctica: Topographic surface, ice thickness, subglacial bed elevation and sediment thickness. *Czech Polar Rep.* **2021**, *11*, 67–85. [CrossRef]
48. Kovalevsky, E.V.; Kharchenko, V.I. Integrated interpretation of marine engineering geological and geophysical data on the principles of expert system technology: Kovalevsky, E V; Kharchenko, V I Geophys ProspectV40, N8, Nov 1992, P909–923. *Int. J. Rock Mech. Min. Sci. Geomech. Abstr.* **1993**, *30*, A234. [CrossRef]
49. Abuamarah, B.A.; Nabawy, B.S.; Shehata, A.M.; Kassem, O.M.; Ghrefat, H. Integrated geological and petrophysical characterization of oligocene deep marine unconventional poor to tight sandstone gas reservoir. *Mar. Pet. Geol.* **2019**, *109*, 868–885. [CrossRef]
50. Veevers, J.J.; Powell, C.M.; Roots, S.R. Review of seafloor spreading around Australia. I. synthesis of the patterns of spreading. *Aust. J. Earth Sci.* **1991**, *38*, 373–389. [CrossRef]
51. Lemenkova, P. Geodynamic setting of Scotia Sea and its effects on geomorphology of South Sandwich Trench, Southern Ocean. *Pol. Polar Res.* **2021**, *42*, 1–23. [CrossRef]
52. Hill, P.J. Seafloor Magnetic Patterns: Seafloor Spreading and Magnetization in the Southern Ocean. *Explor. Geophys.* **1986**, *17*, 40–42. [CrossRef]
53. Nougier, J. *Bibliographie Cartographique des Iles Kerguelen/par Jacques Nougier*; Institut Geographique National: Paris, France, 1968; p. 55. Available online: https://catalogue.nla.gov.au/Record/2259953 (accessed on 10 July 2023).
54. Petit, G.; Petit, G. Edgar Aubert de la Rüe. — Etude géologique et géographique de l'Archipel de Kerguelen. *Rev. D'Écologie* **1933**, *3*, 63–64.
55. Recq, M. Contribution a l'etude de la structure profonde du banc de Crozet (Ocean Indien Austral). *Mar. Geol.* **1989**, *88*, 133–144. [CrossRef]

56. Operto, S. Structure et Origine du Plateau de Kerguelen (Océan Indien Austral): Implications Géodynamique. Modélisation de Données Sismiques Grand-Angle Marines. Ph.D. Thesis, Université Pierre et Marie Curie, Paris, France, 1995.
57. Ramsay, D.C.; Colwell, J.B.; Coffin, M.F.; Davies, H.L.; Hill, P.J.; Pigram, C.J.; Stagg, H.M.J. New findings from the Kerguelen Plateau. *Geology* **1986**, *14*, 589–593. [CrossRef]
58. Breton, T.; Nauret, F.; Pichat, S.; Moine, B.; Moreira, M.; Rose-Koga, E.F.; Auclair, D.; Bosq, C.; Wavrant, L.M. Geochemical heterogeneities within the Crozet hotspot. *Earth Planet. Sci. Lett.* **2013**, *376*, 126–136. [CrossRef]
59. Guglielmi, M. Etude Géophysique du Plateau de Kerguelen. Ph.D. Thesis, Université Louis Pasteur, Strasbourg, France, 1982.
60. Gaina, C.; Müller, R.D.; Brown, B.; Ishihara, T.; Ivanov, S. Breakup and early seafloor spreading between India and Antarctica. *Geophys. J. Int.* **2007**, *170*, 151–169. [CrossRef]
61. Henry, B.; Plessard, C. New palaeomagnetic results from the Kerguelen Islands. *Geophys. J. Int.* **1997**, *128*, 73–83. [CrossRef]
62. Nicolaysen, K.; Frey, F.; Hodges, K.; Weis, D.; Giret, A. 40Ar/39Ar geochronology of flood basalts from the Kerguelen Archipelago, southern Indian Ocean: Implications for Cenozoic eruption rates of the Kerguelen plume. *Earth Planet. Sci. Lett.* **2000**, *174*, 313–328. [CrossRef]
63. Lacroix, A. Les zéolites et les produits siliceux des basaltes de l'archipel de Kerguelen. *Bull. Minéralogie* **1915**, *38*, 134–137. [CrossRef]
64. Doucet, S.; Scoates, J.S.; Weis, D.; Giret, A. Constraining the components of the Kerguelen mantle plume: A Hf-Pb-Sr-Nd isotopic study of picrites and high-MgO basalts from the Kerguelen Archipelago. *Geochem. Geophys. Geosyst.* **2005**, *6*, 1–28. [CrossRef]
65. Li, Z. Structure, origine et évolution du plateau de Kerguelen. Ph.D. Thesis, Université Louis Pasteur, Strasbourg, France, 1988.
66. Aubert de la Rue, E. Aubert de la Rue (Edgar).—Terres françaises: Iles Kerguelen, Crozet, Saint-Paul. *Outre-Mers. Rev. D'histoire* **1931**, *19*, 552–553.
67. Faivre, J.P. Marthe Emmanuel.—La France et l'exploration polaire. I. De Verrazano à la Pêrouse. *Outre-Mers. Rev. D'histoire* **1960**, *47*, 291–292.
68. Dupouy, A. Supplément à l'histoire de Kerguélen (d'après des documents inédits). *Ann. Bretagne Des Pays L'Ouest* **1935**, *42*, 146–169. [CrossRef]
69. Aubert de La Rüe, E. La flore et la faune des îles Kerguelen. *Rev. D'Écologie* **1932**, *2*, 29–51. [CrossRef]
70. Ferchiou, S.; Caza, F.; Villemur, R.; Labonne, J.; St-Pierre, Y. Skin and Blood Microbial Signatures of Sedentary and Migratory Trout (Salmo trutta) of the Kerguelen Islands. *Fishes* **2023**, *8*, 174. [CrossRef]
71. Aubert de La Rüe, E. A propos de la flore des îles Kerguelen. *Rev. D'Écologie* **1932**, *2*, 114–115. [CrossRef]
72. Paulian, P. La vie animale aux îles Kerguelen. *Rev. D'Écologie* **1952**, *6*, 129–138. [CrossRef]
73. Aubert de La Rüe, E. Sur la répartition des grandes colonies de Manchots de la péninsule Courbet (Archipel des Kerguelen). *Rev. D'Écologie* **1953**, *7*, 132–134. [CrossRef]
74. Park, Y.H.; Vivier, F.; Roquet, F.; Kestenare, E. Direct observations of the ACC transport across the Kerguelen Plateau. *Geophys. Res. Lett.* **2009**, *36*. [CrossRef]
75. Damerell, G.M.; Heywood, K.J.; Stevens, D.P. Direct observations of the Antarctic circumpolar current transport on the northern flank of the Kerguelen Plateau. *J. Geophys. Res. Ocean.* **2013**, *118*, 1333–1348. [CrossRef]
76. Mongin, M.; Molina, E.; Trull, T.W. Seasonality and scale of the Kerguelen plateau phytoplankton bloom: A remote sensing and modeling analysis of the influence of natural iron fertilization in the Southern Ocean. *Deep Sea Res. Part II Top. Stud. Oceanogr.* **2008**, *55*, 880–892. [CrossRef]
77. Park, Y.H.; Gasco, N.; Duhamel, G. Slope currents around the Kerguelen Islands from demersal longline fishing records. *Geophys. Res. Lett.* **2008**, *35*, 1–6. [CrossRef]
78. Park, Y.H.; Durand, I.; Kestenare, E.; Rougier, G.; Zhou, M.; d'Ovidio, F.; Cotté, C.; Lee, J.H. Polar Front around the Kerguelen Islands: An up-to-date determination and associated circulation of surface/subsurface waters. *J. Geophys. Res. Ocean.* **2014**, *119*, 6575–6592. [CrossRef]
79. Pollard, R.; Venables, H.; Read, J.; Allen, J. Large-scale circulation around the Crozet Plateau controls an annual phytoplankton bloom in the Crozet Basin. *Deep Sea Res. Part II Top. Stud. Oceanogr.* **2007**, *54*, 1915–1929. [CrossRef]
80. Pauthenet, E.; Roquet, F.; Madec, G.; Guinet, C.; Hindell, M.; McMahon, C.R.; Harcourt, R.; Nerini, D. Seasonal Meandering of the Polar Front Upstream of the Kerguelen Plateau. *Geophys. Res. Lett.* **2018**, *45*, 9774–9781. [CrossRef]
81. Della Penna, A.; Trull, T.W.; Wotherspoon, S.; De Monte, S.; Johnson, C.R.; d'Ovidio, F. Mesoscale Variability of Conditions Favoring an Iron-Induced Diatom Bloom Downstream of the Kerguelen Plateau. *J. Geophys. Res. Ocean.* **2018**, *123*, 3355–3367. [CrossRef]
82. Rietbroek, R.; LeGrand, P.; Wouters, B.; Lemoine, J.M.; Ramillien, G.; Hughes, C.W. Comparison of in situ bottom pressure data with GRACE gravimetry in the Crozet-Kerguelen region. *Geophys. Res. Lett.* **2006**, *33*, 1–4. [CrossRef]
83. Sombetzki-Lengagne, D. La protection de l'environnement dans les Terres australes et antarctiques françaises: Le projet de création d'une réserve naturelle des terres australes. *Rev. Jurid. L'Environ.* **2003**, *28*, 307–317. [CrossRef]
84. Sombetzki-Lengagne, D. Commentaire de la loi n° 2003-347 du 15 avril 2003 relative à la protection de l'environnement en Antarctique. *Rev. Jurid. L'Environ.* **2003**, *28*, 447–460. [CrossRef]
85. Naim-Gesbert, E.; Manouvel, M. Le double visage de la réserve naturelle des Terres australes françaises. *Rev. Jurid. L'Environ.* **2007**, *32*, 445–455. [CrossRef]

86. Wessel, P.; Luis, J.F.; Uieda, L.; Scharroo, R.; Wobbe, F.; Smith, W.H.F.; Tian, D. The Generic Mapping Tools version 6. Available online: https://www.generic-mapping-tools.org/ (accessed on 7 July 2023).
87. Müller, R.D.; Sdrolias, M.; Gaina, C.; Roest, W.R. Age, spreading rates, and spreading asymmetry of the world's ocean crust. *Geochem. Geophys. Geosyst.* **2008**, *9*, 1–19. [CrossRef]
88. NOAA National Centers for Environmental Information. *ETOPO 2022 15 Arc-Second Global Relief Model*; NOAA National Centers for Environmental Information: Asheville, NC, USA, 2022. [CrossRef]
89. Pavlis, N.K.; Holmes, S.A.; Kenyon, S.C.; Factor, J.K. The development and evaluation of the Earth Gravitational Model 2008 (EGM2008). *J. Geophys. Res. Solid Earth* **2012**, *117*, 1–38. [CrossRef]
90. Pavlis, N.K.; Holmes, S.A.; Kenyon, S.C.; Factor, J.K. An Earth Gravitational Model to Degree 2160: EGM2008. *Geophys. Res. Abstr.* **2008**, *10*, EGU2008-A-01891.
91. Braitenberg, C. Gravity. In *Encyclopedia of Geology*, 2nd ed.; Alderton, D., Elias, S.A., Eds.; Academic Press: Oxford, UK, 2021; pp. 706–718. [CrossRef]
92. McNutt, M. Gravity. In *Encyclopedia of Ocean Sciences*, 3rd ed.; Cochran, J.K., Bokuniewicz, H.J., Yager, P.L., Eds.; Academic Press: Oxford, UK, 2019; pp. 20–27. [CrossRef]
93. Condie, K.C. The mantle. In *Earth as an Evolving Planetary System*, 4th ed.; Condie, K.C., Ed.; Academic Press: Oxford, UK, 2022; Chapter 4, pp. 81–125. [CrossRef]
94. Straume, E.O.; Gaina, C.; Medvedev, S.; Hochmuth, K.; Gohl, K.; Whittaker, J.M.; Abdul Fattah, R.; Doornenbal, J.C.; Hopper, J.R. GlobSed: Updated Total Sediment Thickness in the World's Oceans. *Geochem. Geophys. Geosyst.* **2019**, *20*, 1756–1772. [CrossRef]
95. Maus, S.; Barckhausen, U., Berkenbosch, H.; Bournas, N.; Brozena, J.; Childers, V.; Dostaler, F.; Fairhead, J.D.; Finn, C.; von Frese, R.R.B.; et al. EMAG2: A 2–arc min resolution Earth Magnetic Anomaly Grid compiled from satellite, airborne, and marine magnetic measurements. *Geochem. Geophys. Geosyst.* **2009**, *10*, 1–12. [CrossRef]
96. Lesur, V.; Hamoudi, M.; Choi, Y.; Dyment, J.; Thébault, E. Building the second version of the World Digital Magnetic Anomaly Map (WDMAM). *Earth Planets Space* **2016**, *68*, 27. [CrossRef]
97. Sandwell, D.T.; Müller, R.D.; Smith, W.H.F.; Garcia, E.; Francis, R. New global marine gravity model from CryoSat-2 and Jason-1 reveals buried tectonic structure. *Science* **2014**, *346*, 65–67. [CrossRef]
98. Emishaw, L.; Abdelsalam, M.G. Satellite gravity data for mapping lithospheric structure of Precambrian tectonic blocks in Africa: The advantages and limitations. *J. Afr. Earth Sci.* **2023**, *197*, 104775. [CrossRef]
99. Wang, J.; Qi, X.; Luo, K.; Li, Z.; Zhou, R.; Guo, J. Height connection across sea by using satellite altimetry data sets, ellipsoidal heights, astrogeodetic deflections of the vertical, and an Earth Gravity Model. *Geod. Geodyn.* **2023**, *14*, 347–354. [CrossRef]
100. Lehmann, R. Altimetry–gravimetry problems with free vertical datum. *J. Geod.* **2000**, *74*, 327–334. [CrossRef]
101. Nguyen, V.S.; Pham, V.T.; Van Nguyen, L.; Andersen, O.B.; Forsberg, R.; Tien Bui, D. Marine gravity anomaly mapping for the Gulf of Tonkin area (Vietnam) using Cryosat-2 and Saral/AltiKa satellite altimetry data. *Adv. Space Res.* **2020**, *66*, 505–519. [CrossRef]
102. Wessel, P.; Luis, J.F.; Uieda, L.; Scharroo, R.; Wobbe, F.; Smith, W.H.F.; Tian, D. The Generic Mapping Tools Version 6. *Geochem. Geophys. Geosyst.* **2019**, *20*, 5556–5564. [CrossRef]
103. Lemenkova, P. Handling Dataset with Geophysical and Geological Variables on the Bolivian Andes by the GMT Scripts. *Data* **2022**, *7*, 74. [CrossRef]
104. Lemenkova, P. Console-Based Mapping of Mongolia Using GMT Cartographic Scripting Toolset for Processing TerraClimate Data. *Geosciences* **2022**, *12*, 140. [CrossRef]
105. Wessel, P.; Luis, J.F. The GMT/MATLAB Toolbox. *Geochem. Geophys. Geosyst.* **2017**, *18*, 811–823. [CrossRef]
106. Lemenkova, P. Mapping Climate Parameters over the Territory of Botswana Using GMT and Gridded Surface Data from TerraClimate. *ISPRS Int. J. Geo-Inf.* **2022**, *11*, 473. [CrossRef]
107. Neteler, M.; Bowman, M.H.; Landa, M.; Metz, M. GRASS GIS: A multi-purpose open source GIS. *Environ. Model. Softw.* **2012**, *31*, 124–130. [CrossRef]
108. Rocchini, D.; Petras, V.; Petrasova, A.; Chemin, Y.; Ricotta, C.; Frigeri, A.; Landa, M.; Marcantonio, M.; Bastin, L.; Metz, M.; et al. Spatio-ecological complexity measures in GRASS GIS. *Comput. Geosci.* **2017**, *104*, 166–176. [CrossRef]
109. Frigeri, A.; Hare, T.; Neteler, M.; Coradini, A.; Federico, C.; Orosei, R. A working environment for digital planetary data processing and mapping using ISIS and GRASS GIS. *Planet. Space Sci.* **2011**, *59*, 1265–1272. [CrossRef]
110. Lemenkova, P. A GRASS GIS Scripting Framework for Monitoring Changes in the Ephemeral Salt Lakes of Chotts Melrhir and Merouane, Algeria. *Appl. Syst. Innov.* **2023**, *6*, 61. [CrossRef]
111. Sorokine, A. Implementation of a parallel high-performance visualization technique in GRASS GIS. *Comput. Geosci.* **2007**, *33*, 685–695. [CrossRef]
112. Lee, T.R.; Wood, W.T.; Phrampus, B.J. A Machine Learning (kNN) Approach to Predicting Global Seafloor Total Organic Carbon. *Glob. Biogeochem. Cycles* **2019**, *33*, 37–46. [CrossRef]
113. Lemenkova, P. Tanzania Craton, Serengeti Plain and Eastern Rift Valley: Mapping of geospatial data by scripting techniques. *Est. J. Earth Sci.* **2022**, *71*, 61–79. [CrossRef]
114. DeSanto, J.B.; Sandwell, D.T.; Chadwell, C.D. Seafloor geodesy from repeated sidescan sonar surveys. *J. Geophys. Res. Solid Earth* **2016**, *121*, 4800–4813. [CrossRef]

115. Harper, H.; Tozer, B.; Sandwell, D.T.; Hey, R.N. Marine Vertical Gravity Gradients Reveal the Global Distribution and Tectonic Significance of "Seesaw" Ridge Propagation. *J. Geophys. Res. Solid Earth* **2021**, *126*, e2020JB020017. [CrossRef]
116. Fanjat, G.; Camps, P.; Shcherbakov, V.; Barou, F.; Sougrati, M.T.; Perrin, M. Magnetic interactions at the origin of abnormal magnetic fabrics in lava flows: A case study from Kerguelen flood basalts. *Geophys. J. Int.* **2012**, *189*, 815–832. [CrossRef]
117. Weissel, J.K.; Hayes, D.E. Magnetic Anomalies in the Southeast Indian Ocean. In *Antarctica Oceanology II: The Australian–New Zealand Sector*; American Geophysical Union (AGU): Washington, DC, USA, 1978; pp. 165–196. [CrossRef]
118. Jokat, W.; Nogi, Y.; Leinweber, V. New aeromagnetic data from the western Enderby Basin and consequences for Antarctic-India break-up. *Geophys. Res. Lett.* **2010**, *37*, 1–5. [CrossRef]
119. Lemenkova, P. Mapping submarine geomorphology of the Philippine and Mariana trenches by an automated approach using GMT scripts. *Proc. Latv. Acad. Sciences. Sect. B Nat. Exact Appl. Sci.* **2022**, *76*, 258–266. [CrossRef]
120. Lemenkova, P. Geomorphology of the Puerto Rico Trench and Cayman Trough in the Context of the Geological Evolution of the Caribbean Sea. *Ann. Univ. Mariae Curie-Sklodowska Sect. B–Geogr. Geol. Mineral. Petrogr.* **2020**, *75*, 115–141. [CrossRef]
121. Lemenkova, P. Using GMT for 2D and 3D Modeling of the Ryukyu Trench Topography, Pacific Ocean. *Misc. Geogr.* **2021**, *25*, 213–225. [CrossRef]
122. Bradley, L.M.; Frey, H. Constraints on the crustal nature and tectonic history of the Kerguelen Plateau from comparative magnetic modeling using MAGSAT data. *Tectonophysics* **1988**, *145*, 243–251. [CrossRef]
123. Duncan, R.A.; Falloon, T.J.; Quilty, P.G.; Coffin, M.F. Widespread Neogene volcanism on Central Kerguelen Plateau, Southern Indian Ocean. *Aust. J. Earth Sci.* **2016**, *63*, 379–392. [CrossRef]
124. Plenier, G.; Camps, P.; Henry, B.; Ildefonse, B. Determination of flow directions by combining AMS and thin-section analyses: Implications for Oligocene volcanism in the Kerguelen Archipelago (southern Indian Ocean). *Geophys. J. Int.* **2005**, *160*, 63–78. [CrossRef]
125. Houtz, R.E.; Hayes, D.E.; Markl, R.G. Kerguelen Plateau bathymetry, sediment distribution and crustal structure. *Mar. Geol.* **1977**, *25*, 95–130. [CrossRef]
126. Recq, M.; Charvis, P. A seismic refraction survey in the Kerguelen Isles, southern Indian Ocean. *Geophys. J. R. Astron. Soc.* **1986**, *84*, 529–559. [CrossRef]
127. Fröhlich, F.; Wicquart, E. Upper cretaceous and paleogene sediments from the Northern Kerguelen Plateau. *Geo-Mar. Lett.* **1989**, *9*, 127–133. [CrossRef]
128. Duncan, R.A.; Quilty, P.G.; Barling, J.; Fox, J.M. Geological development of Heard Island, Central Kerguelen Plateau. *Aust. J. Earth Sci.* **2016**, *63*, 81–89. [CrossRef]
129. Huang, X.; Wu, S.; De Santis, L.; Wang, G.; Hernández-Molina, F.J. Deep water sedimentary processes in the Enderby Basin (East Antarctic margin) during the Cenozoic. *Basin Res.* **2022**, *34*, 1917–1935. [CrossRef]
130. Robert, C.; Diester-Haass, L.; Chamley, H. Late Eocene-Oligocene oceanographic development at southern high latitudes, from terrigenous and biogenic particles: A comparison of Kerguelen Plateau and Maud Rise, ODP Sites 744 and 689. *Mar. Geol.* **2002**, *191*, 37–54. [CrossRef]
131. Whittaker, J.; Müller, R.; Roest, W.; Wessel, P.; Smith, W.H.F. How supercontinents and superoceans affect seafloor roughness. *Nature* **2008**, *456*, 938–941. [CrossRef]
132. Weis, D.; Frey, F.A.; Schlich, R.; Schaming, M.; Montigny, R.; Damasceno, D.; Mattielli, N.; Nicolaysen, K.E.; Scoates, J.S. Trace of the Kerguelen mantle plume: Evidence from seamounts between the Kerguelen Archipelago and Heard Island, Indian Ocean. *Geochem. Geophys. Geosyst.* **2002**, *3*, 1–27. [CrossRef]
133. Le Pichon, X.; Heirtzler, J.R. Magnetic anomalies in the Indian Ocean and sea-floor spreading. *J. Geophys. Res.* **1968**, *73*, 2101–2117. [CrossRef]
134. Maus, S. Earth Magnetic Anomaly Grid Released. *Eos Trans. Am. Geophys. Union* **2009**, *90*, 239. [CrossRef]
135. Henry, B.; Plenier, G.; Camps, P. Post-emplacement tilting of lava flows inferred from magnetic fabric study: The example of Oligocene lavas in the Jeanne d'Arc Peninsula (Kerguelen Islands). *J. Volcanol. Geotherm. Res.* **2003**, *127*, 153–164. [CrossRef]
136. Xu, G.; Frey, F.A.; Weis, D.; Scoates, J.S.; Giret, A. Flood basalts from Mt. Capitole in the central Kerguelen Archipelago: Insights into the growth of the archipelago and source components contributing to plume-related volcanism. *Geochem. Geophys. Geosyst.* **2007**, *8*, 1–8. [CrossRef]
137. Pettersen, Ø.; Maupin, V. Lithospheric anisotropy on the Kerguelen hotspot track inferred from Rayleigh wave polarisation anomalies. *Geophys. J. Int.* **2002**, *149*, 225–246. [CrossRef]
138. Direen, N.G.; Cohen, B.E.; Maas, R.; Frey, F.A.; Whittaker, J.M.; Coffin, M.F.; Meffre, S.; Halpin, J.A.; Crawford, A.J. Naturaliste Plateau: Constraints on the timing and evolution of the Kerguelen Large Igneous Province and its role in Gondwana breakup. *Aust. J. Earth Sci.* **2017**, *64*, 851–869. [CrossRef]
139. Charvis, P.; Recq, M.; Operto, S.; Brefort, D. Deep structure of the northern Kerguelen Plateau and hotspot-related activity. *Geophys. J. Int.* **1995**, *122*, 899–924. [CrossRef]
140. Recq, M.; Charvis, P. La ride asismique de Kerguelen-Heard—Anomalie du geoide et compensation isostatique. *Mar. Geol.* **1987**, *76*, 301–311. [CrossRef]

141. Coblentz, D.; Chase, C.G.; Karlstrom, K.E.; van Wijk, J. Topography, the geoid, and compensation mechanisms for the southern Rocky Mountains. *Geochem. Geophys. Geosyst.* **2011**, *12*, 1–18. [CrossRef]
142. Sandwell, D.T.; MacKenzie, K.R. Geoid height versus topography for oceanic plateaus and swells. *J. Geophys. Res. Solid Earth* **1989**, *94*, 7403–7418. [CrossRef]

geomatics

MDPI

Review

Introducing Smart Marine Ecosystem-Based Planning (SMEP)—How SMEP Can Drive Marine Spatial Planning Strategy and Its Implementation in Greece

Stilianos Contarinis [1,*], Byron Nakos [1] and Athanasios Pallikaris [2]

[1] Cartography Laboratory, National Technical University of Athens, 15780 Athens, Greece; bnakos@central.ntua.gr
[2] Navigation and Sea Sciences Laboratory, Hellenic Naval Academy, 18539 Piraeus, Greece; palikari@hna.gr
* Correspondence: contarinis@central.ntua.gr; Tel.: +30-210-772-2733

Abstract: This paper introduces smart marine ecosystem-based planning (SMEP), a marine spatial planning (MSP) strategy for more participatory and responsive marine governance by leveraging "smart" digital services. SMEP denotes an iterative MSP process with planning cycles that incorporate continuous data gathering of spatial–temporal natural phenomena and human activities in coastal and marine areas, with ongoing data mining to locate key patterns and trends, to strive for periodic refinement of the MSP output. SMEP aims to adopt an ecosystem-based approach, taking into account both living and nonliving aspects of the marine environment, and making use of all available spatial data at various resolutions. In pursuit of SMEP implementation, the paper examines the current state of the MSP process in Greece and relates its long-term success with the establishment of a marine spatial data infrastructure (MSDI), employing contemporary nautical cartography standards along with hydrospatial data services.

Keywords: smart marine ecosystem-based planning (SMEP); marine spatial planning (MSP); marine spatial data infrastructure (MSDI); marine cadastre; Maritime Limits and Boundaries (IHO S-121); Marine Protected Areas (IHO S-122); maritime service portfolios; hydrospatial data services

Citation: Contarinis, S.; Nakos, B.; Pallikaris, A. Introducing Smart Marine Ecosystem-Based Planning (SMEP)—How SMEP Can Drive Marine Spatial Planning Strategy and Its Implementation in Greece. *Geomatics* **2022**, *2*, 197–220. https://doi.org/10.3390/geomatics2020012

Academic Editors: Giuseppe Masetti, Ian Church, Anand Hiroji and Ove Andersen

Received: 26 April 2022
Accepted: 12 May 2022
Published: 13 May 2022

1. Introduction

Greece is the country with the ninth longest coastline in the world and the third in Europe (https://www.worldatlas.com/articles/countries-in-europe-with-the-longest-coastline.html (accessed on 8 May 2022)) (being approximately 15,000 km at the scale of 1:250,000) [1,2], with approximately 6000 islands and islets (https://www.visitgreece.gr/islands/ (accessed on 8 May 2022)), offering a highly diversified landscape. Based on OECD data [3], in the coastal zone of the country are concentrated almost 80% of industrial activity, 90% of tourism and leisure, 35% of rural land, and a significant portion of basic infrastructure (ports, airports, roads, electricity, telecommunications). However, despite its significant maritime wealth, Greece still lacks national spatial planning for the development of maritime activities.

The objectives of this paper are initially to have a short background on MSP in Section 2, highlighting ecological concerns and to describe the international and regional marine governance frameworks in Sections 3 and 4. In Section 5, the current situation in the MSP process in Greece is presented based on the recent legislation. In Section 6, the concept for smart marine ecosystem-based planning is introduced to accommodate a holistic view the constantly evolving "smart" needs of the marine areas. In Section 7, the possible next steps in the marine cadastre implementation in Greece is discussed facilitated by a marine spatial data infrastructure (MSDI), highlighting the use of relevant data model specifications of the International Hydrographic Organisation (IHO) for Maritime Limits and Boundaries and Marine Protected Areas. In addition, public cloud services (G-Cloud) are used as platforms to host the MSDI and serve the various stakeholders. Finally, in Section 8, the developments for the marine spatial planning (MSP) in Greece is discussed in the context of the post COVID-19 state's plan for reforms.

2. MSP Background

2.1. Ecological Concerns

Human activities have frequently resulted in measures to protect marine areas. For example, the Great Barrier Reef in Australia was conserved in the 1970s from the threat of offshore oil and gas extraction [4]. Likewise, in the Mediterranean Sea, the marine regions in the Ionian and Cretan seas (about 56,000 km^2), which have been granted as concessions to the oil and gas industry for hydrocarbon exploration, require protection [5]. From north of Corfu to southern Crete, the area coincides with the southwest Hellenic Trench, an important habitat and marine biodiversity hotspot of global ecological significance.

2.1.1. The Hellenic Trench

The deepest point in the Mediterranean Sea is located in the southwest of Greece, 62 miles off the coast of Cape Tainaros, and has a depth of 5120 m (https://iskra.gr/%CF%84%CE%B1-%CE%BC%CE%B5%CE%B3%CE%B1%CE%BB%CF%8D%CF%84%CE%B5%CF%81%CE%B1-%CE%B2%CE%AC%CE%B8%CE%B7-%CE%B8%CE%B1%CE%BB%CE%B1%CF%83%CF%83%CF%8E%CE%BD-%CF%83%CF%84%CE%B7%CE%BD-%CE%B5%CE%BB%CE%BB%CE%AC/ (accessed on 8 May 2022)). It is about half the depth of the Mariana Trench, the deepest point on Earth, with a depth of 11,035 m, located between Indonesia and Japan. The second deepest point in the Mediterranean Sea is the deep of Rhodes, 14 miles east of the rock Paximada, with a maximum depth of 4710 m, and the third one is the deep southwest of Karpathos Island, with a maximum depth of 3294 m.

All three are part of the Hellenic Trench (Figure 1), a lengthy bathymetric feature consisting of a continuous steep continental seaward slope, bounded by offshore linear trenches, troughs, and basins. The area is a habitat for the endangered sperm whale (https://www.arion.org.gr/mammal/sperm-whale/ (accessed on 8 May 2022)) that includes some 200–250 animals threatened by potential ship-strikes and seismic blasts [6]. Because of the deep water and the seabed features, the Hellenic Trench is a shelter for deep-diving marine animals such as beaked whales, fin whales, several types of dolphins, Mediterranean monk seals, and sea turtles.

2.1.2. The Ionian Archipelago

In the adjacent Northern Ionian Archipelago, the population of common dolphins was substantially decreased a few decades ago, but the latest research has indicated that they are likely to be spread across the whole archipelago. The region also has one of the most important populations of Mediterranean monk seals (Monachus monachus), accounting for around 7% of the global population, and the breeding sites for pupping in the region have been systematically researched [7].

2.2. Important Marine Mammal Areas

Marine mammal specialists from across the globe have recognized the Hellenic Trench and the Ionian Sea Archipelago as "important marine mammal areas", while the Hellenic Trench has been also proposed as a marine protected area [5]. Moreover, the agreement for the Conservation of Cetaceans in the Black Sea and the Mediterranean Sea, known as ACCOBAMS (https://accobams.org/conservations-action/protected-areas/ (accessed on 8 May 2022)), has also acknowledged the ecological importance of the Hellenic Trench. However, just a small portion of Greece's southwest coast has been designated for protection thus far.

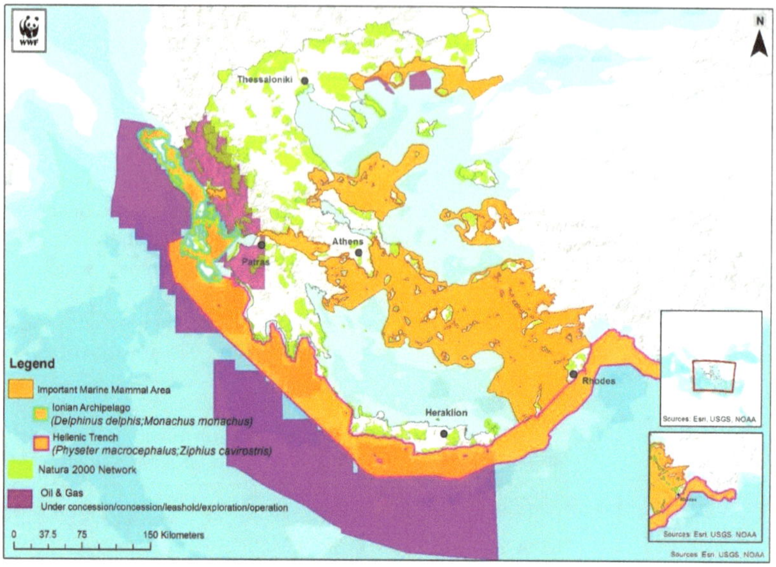

Figure 1. The Hellenic Trench (min. lat. 34°8′ N, min. long. 20°15′ E, max. lat. 37°15′ N, max. long. 28°0′ E) (https://www.marineregions.org/gazetteer.php?p=details&id=3347 (accessed on 8 May 2022)). Important marine mammal area. Reprinted with permission from Ref. [5]; Copyright 2019 WWF Greece.

2.3. *Marine Spatial Planning*

Marine spatial planning (MSP) is already used in over 60 countries worldwide to identify and resolve conflicts between competing uses of marine space in conjunction with ocean environment conservation programs. MSP is typically driven by the need to identify possible places for new uses in crowded waterways, such as energy extraction and production, and to reduce spatial and temporal conflicts between uses, as well as between demand and environmental protection [8].

MSP and Integrated Coastal Zone Management

The MSP process is related to the objectives of Integrated Coastal Zone Management (ICZM (http://www.coastalwiki.org/wiki/Integrated_Coastal_Zone_Management_(ICZM) (accessed on 8 May 2022))), which is defined by the European Union [9] as "a dynamic, multidisciplinary and iterative process to promote sustainable management of coastal zones. It covers the full cycle of information collection, planning, decision making, management and monitoring of implementation". Moreover, according to the Baltic Marine Environment Protection Commission, also known as Helsinki Commission (HELCOM), there is a need to clarify the role of ICZM, as due to adoption of the MSP, many countries do not implement ICZM as a separate activity [10]. HELCOM is an international body that oversees the Baltic Sea Area Convention for the Protection of the Marine Environment, which suggests ten principles [11] that provide helpful direction for establishing greater cohesion for the region's growth, where the sixth principle is related to the quality of data available.

3. International Governance Framework for the Sea

3.1. *United Nations Convention for the Law of the Sea (UNCLOS)*

The United Nations, in the context of its founding principles and in order to contribute to the preservation of peace and justice in all the countries, confirmed at the Geneva Conferences in 1958 and 1960 that there should be an acceptable convention for the law of the sea. The complete version of the United Nations Convention on the Law of the Sea (UNCLOS) was presented in 1982 in Montego Bay, Jamaica (https://treaties.un.or

g/Pages/Treaties.aspx?id=21&subid=0&lang=en&clang=_en (accessed on 8 May 2022)), entered into force in 1994, and is today the world's most recognized maritime law regime. The central idea of the UNCLOS convention is that maritime problems are interconnected and should be tackled as a whole. Greece ratified the Convention on the Law of the Sea (https://www.un.org/depts/los/convention_agreements/convention_declarations.htm (accessed on 8 May 2022)) in June 1995 (Law 2332/1995, Government Gazette 136/A/23-06-1995) [3]. The provisions of the UN convention apply to all areas of marine affairs, including the organization and development of productive activities and the emergence of marine entrepreneurship.

3.2. Sea Zones in Accordance with UNCLOS Provisions

According to UNCLOS [12] the following maritime zones can be distinguished (Figure 2) in marine areas:

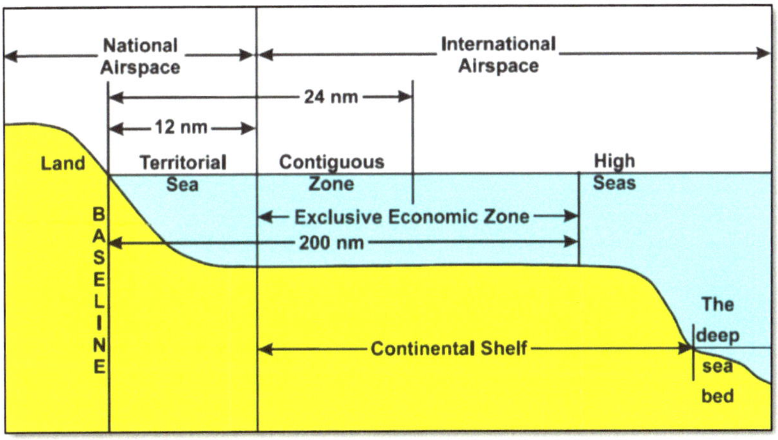

Figure 2. Maritime zones. Reproduced from footnote link (https://sites.tufts.edu/lawofthesea/chapter-two/ (accessed on 8 May 2022)), with permission of Tufts University.

- **Internal Waters** include the sum of the stagnant or flowing surface water and groundwater located on the landward side relative to the territorial sea baseline. Inland waters are dominated by the coastal state and ships from other countries are not allowed to freely pass.

- **Territorial Sea** (or coastal zone) is the zone extending over a range of up to 12 nautical miles from the shoreline (baseline), involving the water column, bottom, subsoil, and airspace, within which states are free to impose any rule of law, regulate any use, and exploit any resource. In the territorial waters, the right of "innocent passage" of ships and aircraft is permitted (continuous, fast transit that does not disturb the peace and security of the coastal state).

- **Contiguous Zone** is the zone having an internal boundary outside the territorial waters and an external boundary of up to 24 nm from the baseline of territorial waters. In this border zone, the coastal state does not have complete authority, but has the necessary control to prevent specific infringements concerning its national legislation in the field of health, customs, migration, and economic matters.

- **Exclusive Economic Zone** (EEZ) is the seabed and its subsoil extending beyond territorial waters to (potentially) a distance of 200 nm from the baseline (Figure 3). In the EEZ, the state does not exercise full sovereignty but sovereign rights. The EEZ regime covers all natural resources, whether living or not, as well as their economic exploitation, research, and environmental protection activities. In the case that the EEZs touch each other, it is up to the countries that are demanding them to jointly define maritime borders.

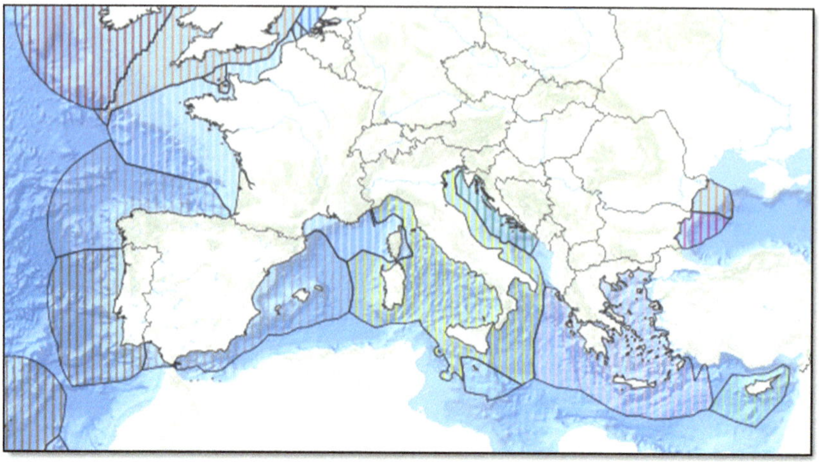

Figure 3. Borders of EU countries of the Mediterranean Sea from the view of the exclusive economic zones. Reproduced from https://emodnet.ec.europa.eu/en/map-week-%E2%80%93-exclusive-economic-zones (accessed on 8 May 2022)), with permission of EMODNET.

According to UNCLOS, an EEZ has only the inhabited islands, resulting from the presence of inhabitants, a lighthouse guardian, farm animals, cultivated land, and anything else that proves economic activity. However, for all other islands and rocks, the article of 12 nautical miles of the coastal zone is normally applied. A state's rights in the EEZ are only created after a declaration of its sovereign rights to the UN. The EEZ regime explicitly permits the construction and use of artificial islands, installations and other structures, scientific research, and the protection and preservation of the marine environment. Within the EEZ of one state, all states have the right to navigate, flight, cable and pipe laying, and other uses in accordance with the international law of the sea.

- **Continental Shelf** is the seafloor and the subsurface extending beyond territorial waters up to (potentially) 200 nautical miles (and up to the outer boundary of the continental shelf if it extends over the 200 nm). In the continental shelf, the state does not exercise full authority but has the rights to extract and make use of the natural resources. Its meaning is weakened because it overlaps with that of the exclusive economic zone (EEZ).

- **High Seas** are located beyond the boundary of the continental shelf and where the premise of innocent passage, fishing, cable laying, and scientific research pipelines applies.

3.3. Delimitation of Maritime Zones

According to this article [13], the UNCLOS mandated delimitation of marine zones is a driver of economic development, a management tool for the marine environment, and the foundation for spatial planning. Maritime regions and limits determine the borders of coastal nations, and their precise demarcation and cartographic portrayal is obligatory for every state. Even though the UN convention is a legislative document, its execution is purely technical, requiring scientific and practical knowledge of geoinformatics for those engaged.

Maritime Zones in Eastern Mediterranean

The delimitation of maritime zones in the Mediterranean Sea, and in particular in the Eastern Mediterranean, is well known and ambiguous, because as well as geographical, it is also political. In Greece today, there are declared territorial waters for the marine area at 6 nm from the baseline in accordance with Law No. 230/1936 (Government Gazette A-450/13-10-1936) (and 10 nm for airspace). Greece has not yet designated an EEZ with any neighboring country, although it has the right to do so in accordance with international maritime law and international law.

Disputes between Greece and Turkey over the Aegean continental shelf stretch back to 1973 (https://www.mfa.gr/en/issues-of-greek-turkish-relations/ (accessed on 8 May 2022)), when the Turkish Government released a permit to the national petroleum corporation to conduct research on the Greek continental shelf in the Eastern Aegean. Since then, Turkey's persistent attempts to infringe on Greece's sovereign rights to the continental shelf have become a major cause of conflict in the bilateral relations between the two nations. This article [14] takes into account the positions of the two countries as they have been formally articulated. The Greek perspective is that territorial waters demarcation should be ruled by the average line principle, while the Turkish perspective is that demarcation should be conducted in such a manner that it creates an equitable result, that does not essentially authorize islands to entire maritime zones.

3.4. The UNEP Regional Seas Programme

The concept of regional seas is used to identify and describe policies aimed at establishing and supporting multilateral, transnational cooperation networks in maritime clusters addressing common environmental issues. The ultimate goal is to create networks and partnerships between states to promote sustainability and identify the benefits of cooperative management of marine areas. The United Nations plays a central role in supporting such cooperation networks, and from the early 1970s has been trying to tackle integrated marine environmental problems through the Regional Seas Programme (RSP) (https://www.unep.org/explore-topics/oceans-seas/what-we-do/regional-seas-programme (accessed on 8 May 2022)), currently developed for 13 marine areas. This program is being implemented in the context of the United Nations Environment Programme (UNEP) (https://www.unep.org/ (accessed on 8 May 2022)), a central body for establishing the global agenda for the environment.

The Mediterranean Action Plan

Greece and the Mediterranean coastal countries are included in the Mediterranean Action Plan (MAP), which is the first project implemented under the UN Regional Seas Program. In 1976, the Mediterranean Action Plan was ratified by fourteen Mediterranean countries and has since become institutionalized through the Barcelona Convention (https://ec.europa.eu/environment/marine/international-cooperation/regional-sea-conventions/barcelona-convention/index_en.htm (accessed on 8 May 2022)). The Mediterranean Action Plan was amended in 1995 (MAPII) and the revised Barcelona Convention has been in force ever since. The headquarters of the Mediterranean Action Plan—Coordination Unit is in Athens, is responsible for the Barcelona Convention Secretariat, and develops general strategies, the latest being the Mid-Term Strategy (MTS) (https://wedocs.unep.org/bitstream/handle/20.500.11822/6071/16ig22_28_22_01_eng.pdf (accessed on 8 May 2022)) 2016–2021. The goal of this strategy is "a healthy Mediterranean with marine and coastal ecosystems that are productive and biologically diverse contributing to sustainable development for the benefit of present and future generations".

4. European Governance Framework for the Sea

4.1. Directive on the Conservation of Natural Habitats

The EU Directive on the Conservation of Natural Habitats and of Wild Fauna and Flora (92/43/EEC) sets out the various procedures and commitments regarding the management of nature conservation in Natura 2000 sites (Figure 4), habitats, and species therein, including marine areas. These provisions have been transposed into the Greek Law 4014/2011. In fact, in the recent joint ministerial decision where the list of Natura sites in Greece were revised, the new areas that are part of the network mainly concern the marine area. The total marine area covers about 22% of national territorial waters, well above the 6.12% of territorial waters previously integrated.

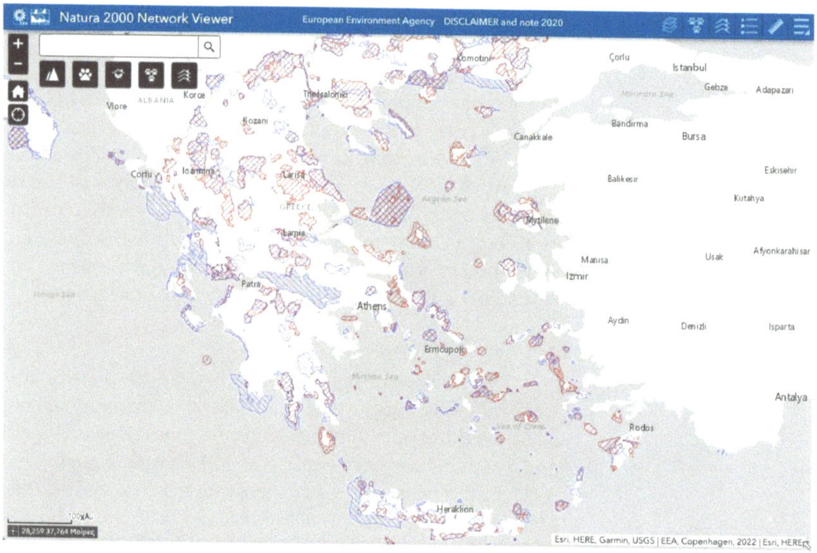

Figure 4. Natura 2000 Network Viewer. Reproduced from https://natura2000.eea.europa.eu/ (accessed on 8 May 2021).

4.2. *Integrated Maritime Policy*

The European Union (EU), recognizing the significant growth opportunity and dynamics of the seas, introduced the Integrated Maritime Policy (IMP), in order to organize and develop maritime cross-sections [15]. Due to the maritime space having many peculiarities, including demarcation of knowledge and available data for geomorphology and environmental status, the pursuit of a new policy was a complex endeavor. The IMP was formally introduced in 2007, laying the foundations for a new European space development strategy, and to date, significant developments have been recorded. IMP defines MSP as a cross-sector policy instrument enabling public authorities and stakeholders to adopt a coordinated, integrated, and cross-border approach. The Marine Strategy Framework Directive (2008/56/EU) is the environmental pylon of the integrated policy that immediately followed and was incorporated in Greek law 3893/2011 (Government Gazette 144/A/17-6-2011).

4.2.1. Blue Growth Strategy

In 2012, EU introduced a long-term strategic planning focused towards long-term sustainability of the marine economic sectors. The strategy was introduced with the communication "Blue Growth: opportunities for marine and maritime sustainable growth" [16], having as the main idea that the oceans are vital parameters for the development of the European economy, with comparative advantages in the fields of innovation and employment. Through the Blue Growth initiative, IMP could achieve the objectives set under the Europe 2020 strategy [17] for smart, sustainable, and inclusive growth. The strategy consists of three components:

(a) Developing the marine sectors with the promise for long-term jobs creation and expansion.
(b) Providing knowledge, regulatory stability, and confidence in the ocean economy.
(c) Implementing sea basin policies to ensure states' collaboration.

4.2.2. Data and Knowledge about the Sea

In August 2012, the green paper "Marine Knowledge 2020: from seabed mapping to ocean forecasting" was published [18]. The Knowledge of the Sea 2020 Strategy, in addition to providing a comprehensive methodological framework and guidelines for marine data, aimed to make a real contribution to knowledge through the gathering of data from various sources. The ultimate goal was to facilitate the access of all interested public authorities and research organizations to marine data and to exploit the development of the Union's maritime policy priorities for exploring new areas of activity.

On this direction, the EU has undertaken the recording of marine data, at Union level, through the European Maritime Observation Network (EMODnet). This network is in its third phase of implementation by recording data across the marine area (surface, water column, bottom, and subsurface) with the participation of 150 research organizations (see Figure 5). Through seven thematic websites (Figure 6), EMODnet facilitates access to maritime data. As outlined in [19], EMODnet's goals are to:

- Enhance efficiency in all operations that use marine data by minimizing data recollection and the expenses for gathering.
- Boost competitiveness and creativity in existing and emerging maritime industries.
- Decrease ambiguity in our understanding of the oceans and increase our ability to forecast.

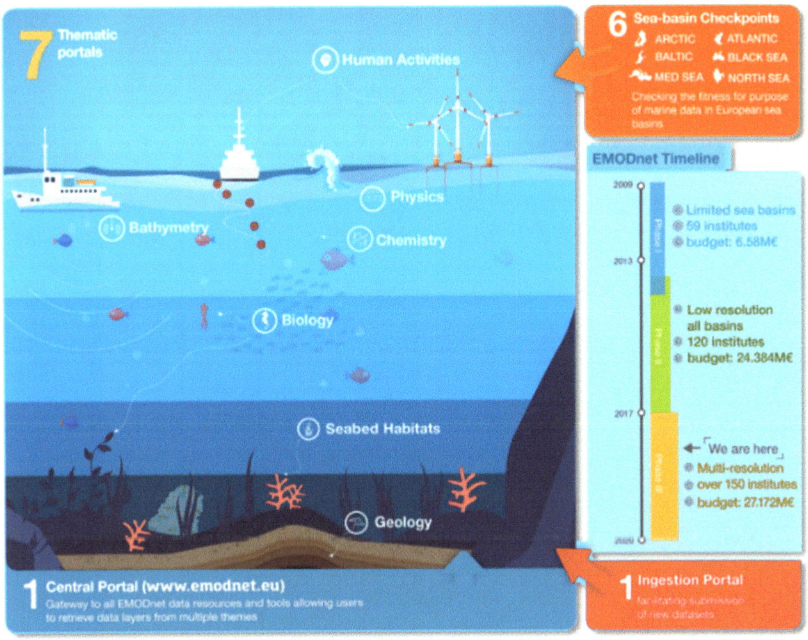

Figure 5. EMODnet infographic. Reproduced from https://emodnet.eu/sites/emodnet.eu/files/public/image_news/EMODnet_in_a_Nutshell.JPG (accessed on 8 May 2022).

4.3. Marine Spatial Planning Directive

In 2014, the EU adopted the Directive 2014/89/EU [20] on the creation of a common framework for maritime planning in Europe. The Directive was incorporated as national law by the end of 2016, and all Member States should have maritime spatial planning studies in place by March 2021.

According to the preamble to Directive 2014/89/EU:

a. The rising interest for marine space for various marine activities, such as energy plants, oil and gas extraction, shipping, fishing, biodiversity conservation, tourism, and underwater cultural heritage, in conjunction with the numerous demands on coastal resources necessitate a holistic approach to planning and management of marine domain.

b. Adopting an ecosystem-based approach will aid in the long-term development and expansion of marine and coastal economies and the responsible use of coastal and marine resources.
c. To encourage the sustained coexisting of activities and, when applicable, the proper placement of complementary uses in the marine region, a framework is needed that typically includes the acceptance and execution by Member States of marine spatial planning outcomes in appropriate charts.

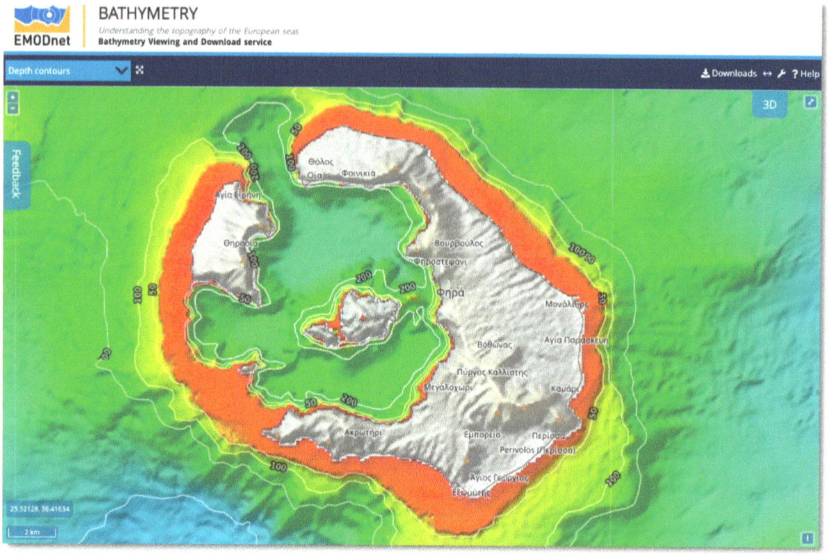

Figure 6. EMODnet Bathymetry portal. Reproduced from https://portal.emodnet-bathymetry.eu (accessed on 8 May 2022).

Environmental Impact Assessments

Special reference is made in the preamble of EU Directive 2014/89 to Directive 2001/42/EU, establishing the environmental impact assessment as a significant instrument for incorporating environmental concerns into project planning and programs approval, and it is noted that where maritime spatial plans:

a. Have a serious environmental impact, they are subject to Directive 2001/42/EU.
b. Incorporate Natura 2000 sites and to prevent overlap, the environmental impact assessment shall be supplemented with the criteria of Article 6 of Directive 1992/43/ EU.

4.4. MSP Plans in Europe

In order to ensure that maritime spatial planning is based on reliable data, Member States must utilize the best possible information and data, enabling parties to share data and information, as well as using existing data and data collecting mechanisms, such as those established under the "Knowledge for the Sea 2020" initiative and Directive 2007/2/EC. MSP plans progress in Europe is monitored closely at msp-platform.eu, and below, the main ones are reviewed.

4.4.1. Germany

In Germany (https://maritime-spatial-planning.ec.europa.eu/countries/germany/ (accessed on 8 May 2022)), MSP is based on the Federal Land Use Planning Act, which has been extended to the exclusive economic zone. The German Länder prepare spatial plans for the territorial sea (out to 12 nautical miles). The plans in Germany are both legislative and effective. The North Sea federal plan entered into force in September 2009, while the Baltic Sea federal plan went into effect in December 2009.

Maritime planning in Germany is based on strategic guidelines, such as:

- Securing and enhancing maritime traffic.
- Enhancing maritime business activities with integrated spatial organization and efficient use of space.
- Encouraging the use of offshore wind energy in accordance with the federal government's sustainability plan.
- Long-term safeguarding and use of rotation mechanisms in case of problems and giving priority to specific uses.
- Ensuring natural resources for tackling ecosystem disturbances and marine pollution.

Germany uses three types of zones to implement its spatial plans, that include:

- Priority areas where one use (for instance, shipping, energy, etc.) takes precedence over all the other uses in the region.
- Reserve areas in which one use is given particular attention to benchmarking with other spatial planning activities.
- Marine protected areas where sustainable measures are applied to the marine environment.

4.4.2. United Kingdom

In England, the UK Government has delegated the responsibilities of marine planning to the Maritime Management Organization (MMO) (https://www.gov.uk/government/collections/marine-planning-in-england (accessed 8 May 2022)), for developing marine projects in coastal and offshore areas. Eleven maritime areas have been established around the coast of England and each area will be covered by a maritime plan with a long-term (20 year) distribution of activities. The plans are being developed on an ongoing basis for the various regions, with the aim of fully covering the English coastal and offshore areas by 2021.

4.4.3. France

In France (https://maritime-spatial-planning.ec.europa.eu/countries/france (accessed 11 July 2021)), there are currently no official maritime spatial plans in existence. The coasts of France are governed by four Interregional Directorates (Direction InterRégionale de la Mer—DIRM) including East Channel–North Sea, North Atlantic–West Channel, South Atlantic, and Mediterranean Sea. Plans are developed for each of the four coasts under the jurisdiction of the préfets coordonnateurs, the préfet de region appointed for that purpose, and the préfet marine. In 2016, the MSP Directive was transposed into French legislation through the entry into force of article123 of law n° 2016-1087 for the second *"reconquest of biodiversity, nature and landscapes"*. Legislation amends the French Environmental Code by introducing the concept of marine spatial planning, which is described as *"the process by which the State defines and organises human activities at sea in an ecological, economic and social perspective. It does not apply to activities related to defense or national security"*.

5. Marine Spatial Planning in Greece

5.1. MSP Law (4546/2018)—Marine Spatial Strategy

Greece incorporated the EU Directive with Law 4546/2018 (Government Gazette 101 A'12.6.2018) (https://www.e-nomothesia.gr/kat-periballon/prostasia-thalassiou-periballontos/nomos-4546-2018-phek-101a-12-6-2018.html (accessed on 8 May 2022)). The first chapter, which includes Articles 1–4, defines the purpose of its provisions, their scope, the definitions, and the objectives of maritime spatial planning. The second chapter of the law includes Articles 5–12, where:

- Article 5 provides for the procedure for the establishment and implementation of maritime spatial planning, which specifies that maritime spatial planning shall be completed as soon as possible and by 31 March 2021 at the latest and stipulates that the maritime spatial planning includes the national spatial planning strategy for sea space and the marine spatial plans.

- Article 6 defines the structure of the marine spatial planning and specifies that the national spatial strategy for the marine area is part of the national spatial strategy of article 3 of Law 4447/2016.
- Article 7 sets out the minimum requirements for maritime spatial planning, while Article 8 defines its content. Article 9 contains provisions for public consultation and public participation, and then Article 10 provides for issues related to the use and exchange of data.
- Articles 11 and 12 then provide for co-operation with Member States and third countries, respectively, while Article 13 sets out the obligation to monitor marine spatial planning.

5.1.1. MSP Authority

In the third chapter of the law, article 13, the Minister of Environment and Energy is defined as authority for maritime spatial planning and the responsibilities are specified. Finally, the fourth chapter of the law provides for transitional provisions for the approval of maritime spatial planning.

Other key points set out in the preamble of Law 4546/2018 are as follows [21]:

- In the Greek maritime and coastal areas, as well as in Europe, human activities, the effects of climate change, and natural disasters, as well as natural coastal transformations, can have serious economic, social, and environmental impacts.
- Marine spatial planning (MSP) is the public process of analyzing and planning the distribution of human activities in marine areas to achieve economic, environmental, and social objectives.
- Through the preparation of plans, the main purpose of maritime spatial planning is to promote sustainable development and determine the utilization of marine space for different uses, as well as the management of their uses and conflicts.
- It is particularly emphasized that another approach that forms part of the Integrated Maritime Policy of the European Union and is directly linked to MSP is the Integrated Coastal Zone Management (ICZM).
- The seamless link between marine and coastal areas requires the coordination and integration of maritime spatial plans and integrated coastal zone management strategies, in order to ensure the sustainable use of maritime space and the management of coastal zones, taking into consideration social, economic, and environmental factors.
- Cooperation between Member States, as well as with third countries, in the maritime areas concerned, in accordance with international law and conventions, in particular the provisions of the UNCLOS Convention, is essential when designing and implementing maritime spatial planning.

5.1.2. Planning for Coastal Land and for the Sea

According to the Regulatory Impact Assessment Report of Law 4546/2018, the law incorporates maritime spatial planning into the existing spatial planning system and seeks coordination and coherence between its spatial planning and marine spatial planning, noting for sectoral policies that:

Economy: ensures the sustainable development of all maritime economic sectors, as it is the process of organizing human activities in maritime and coastal areas in order to achieve the synthesis of ecological, environmental, economic, social, and cultural parameters.

Society: will contribute to improving the services provided and legal certainty for those active in the maritime economy as well as enhancing and securing jobs by organizing maritime activities and uses.

Environment: on the one hand, its ecosystem approach, and on the other hand, ensuring that the coexistence of uses and activities will minimize the impact on the natural and cultural environment.

Public Administration and Justice: the coordination of sectoral policies and the response caused by the lack of integrated spatial planning will facilitate the role of public administration and justice in addressing the spatial effects of the conflicting sectoral provisions.

According to HELCOM [11], land and sea spatial planning should be inextricably linked, consistent, and supportive of one another. To the greatest extent possible, legal systems governing land and sea spatial planning should be harmonized in order to achieve governance systems that are equally open to dealing with land and sea spatial challenges, problems, and opportunities, as well as to create synergies. Moreover, synergies with ICZM should be strengthened in a cross-border setting.

A key added value resulting from the integration of maritime spatial plans and integrated coastal zone management strategies is the strengthening of land–sea connectivity by requiring coherence between maritime spatial planning and integrated coastal planning. Both MSP and ICZM are complementary management tools under IMP [22]. Together, they support a more integrated decision-making process that coordinates, potentially, competing sectoral policies, thereby contributing to the achievement and coherence of objectives and measures in the context of other relevant policies, including energy, the environment, maritime transport, tourism, and fishing. Taken together, they will improve the spatial planning and management of the intermediate zone between land and sea and allow for better coordination of maritime and coastal activities, which may subsequently lead to significant financial and economic benefits for investors as well as reduction in coordination costs.

5.1.3. Planning on Multiple Scales

According to [23], MSP and the regulation of marine uses should be implemented on multiple scales (local, regional, and national), and, especially at the local scale, should not be confined to the marine space, but should include land space, as in the ICZM. For the better implementation of the ecosystem approach, MSP should not be confined to the territorial waters of a coastal state but should as far as possible exhaust EEZ boundaries to include sets of ecosystems (and not just subdivisions thereof).

6. Smart Marine Ecosystem-Based Planning (SMEP)

6.1. Smart Marine Ecosystem

Smart technologies can provide numerous opportunities, when implemented in a holistic, methodical manner that is based not just on technology, but also on successful cooperation, partnerships, and a coherent regulatory environment. At the 2019 Smart4Sea conference, it was highlighted that [24] as the world becomes more connected, the opportunities offered by smart technologies will reach new levels of collaboration and knowledge sharing across all maritime stakeholders. Marine operators can gain greater efficiency by employing smart technology, leading in higher revenues, as well as allowing sustainable communities. Every year, billions are lost due to inefficiencies in the maritime sector, including vessel and port operational accidents. The shipping industry also contributes to climate change [25], which affects everyone. The goal of marine administration should be to pave the way for the transition to a smart marine ecosystem. Smart marine ecosystems, similar to smart cities, are areas where conventional services are made more effective via the use of digital and communication technology for the benefit of both people and the marine population.

Smart Marine Ecosystem-Based Planning Strategy

In this regard, smart marine ecosystem-based planning (SMEP) is proposed as an MSP strategy for going beyond the conventional use of information and communication technologies (ICT) for marine sustainable development, across the three marine information domains [26] (Figure 7), being the marine environment protection, the maritime safety, and the marine spatial planning. SMEP's holistic approach across the domains entails a more participatory and responsive maritime administration for safer marine environments and for better addressing the requirements of the marine population and stakeholders.

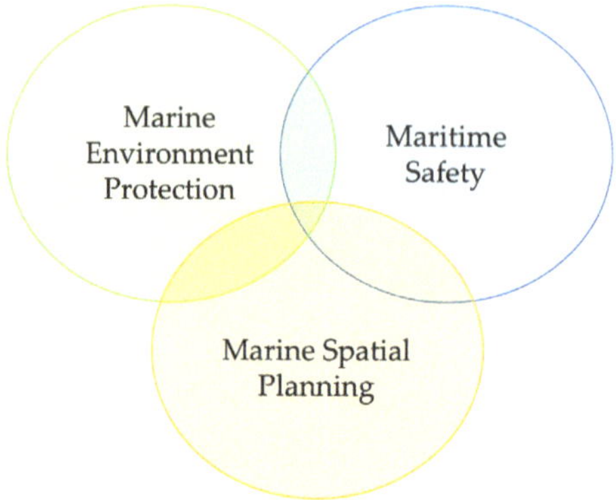

Figure 7. Marine information domains. Reproduced with permission from reference [24]. Copyright 2020, NTUA Cartography Laboratory.

SMEP strategy aims to achieve:

- Improved aquaculture and coastal waste recycling infrastructure.
- More efficient methods of protecting and utilizing maritime space.
- Increased marine life by using more sustainable integrated solutions.
- Smarter maritime transportation networks.
- Solving policy issues in sectors such as fisheries, marine, energy, and information and communication technology.

SMEP is about making the best use of resources while having the least amount of influence on the environment and the highest level of safety. It aids in averting conflicts and optimizing operations by using data from different sensors and data analytics of accessible information, such as shipping routes and weather. SMEP may also be viewed as a transactional platform for driving value and process improvement for both the ecosystem and marine populations. It is focused on the collaboration of the community, industry, and other relevant parties in developing new solutions and participating in maritime space administration. Its objectives are to create and deploy integrated smart marine solutions to enable networking, collaborations, and information sharing, with an emphasis on the confluence of environment, information technology, and maritime.

6.2. Smart Hydrospatial Data Services

In the following paragraphs, "smart" hydrospatial data services related to SMEP are briefly discussed, highlighting the diversity and the potential this planning strategy can achieve (Figure 8).

6.2.1. Smart Ports

With shifting global trade needs, ships that are becoming bigger, products that need to be moved quicker, and geopolitical tensions that are generating new issues for ports all over the globe, the need to adapt and become "smart" is a requirement today. Three main sources of waste have been identified in the maritime sector [24]: excess supply, fuel economy, and wait period at ports and other high-traffic locations. On the other hand, there are four primary forces that can overcome them:

- Pooled ability for increasing capacity utilization and lowering costs.
- Data science, which is connected to digitization, for operational optimization.
- Intelligent vessels for automated and improved processes, as well as performance management.

- Automation to improve port operational effectiveness.

A smart port is one that improves its performance using automation and innovative technologies such as big data, smart sensors, blockchain and artificial intelligence. Although maritime industry is often regarded as conservative, there are solutions based on emerging technologies that alter this perception, steering the whole industry toward a more interconnected era. Vessel traffic control systems have been implemented in the maritime sector to improve port operations and effectively control vessel traffic in ports, harbors, and coastal regions. In this context also, this paper [27] tries to create a smart port paradigm and a quantifiable indicator, the smart port index (SPI), that ports may utilize to enhance their reliability and sustainable development.

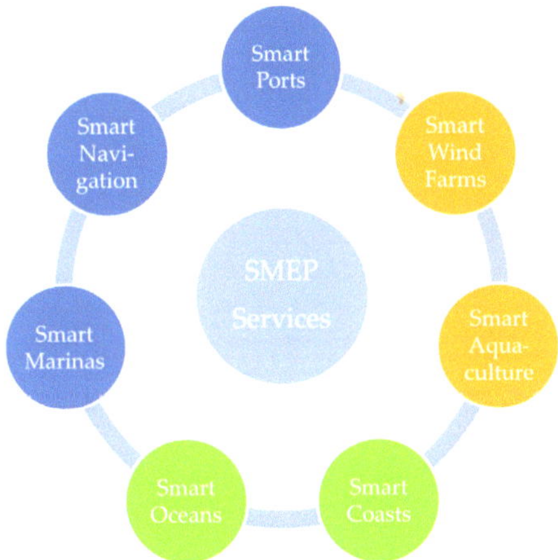

Figure 8. Smart hydrospatial data (SMEP) services across the marine information domains.

6.2.2. Smart Offshore Wind Farms

Related to offshore wind farms (OWFs), research [28] demonstrates the establishment of a spatial decision support system (SDSS) for designating areas for farm siting in accordance with MSP requirements, exclusion criteria, and assessment of environmental and socioeconomic criteria. As proposed by the author, it is critical to have the capability to provide a comprehensive management toolkit for renewable energy policymakers and stakeholders, incorporating a set of quantitative research and spatial–economic models. Moreover, based on characteristics of offshore wind farms and meteorology, as well as the plan construct a run lifecycle process, this paper studied [29] the placement of a smart offshore windfarm that reduces installation and operational costs, increases production capacity, increases equipment life, and ensures personnel safety.

6.2.3. Smart Aquaculture

Aquaculture, especially in developing countries, is one of the most important components in fulfilling the human community's food demand [30]. Year after year, many fish farmers suffer unfathomable losses as a result of unforeseeable circumstances, managerial and operational mistakes, or technical breakdowns. The water quality inside the tanks or around the cages is crucial for optimizing feeding strategy, ensuring growth, reducing mortality, and stimulating reproduction. Smart aquaculture management requires technological investment and reliable control of several environment uncertainties. In that spectrum, water quality management systems are integrated solutions for water quality monitoring, decision support, and process automation based on real-time data.

6.2.4. Smart Coasts

Blue Flag accreditation is an international program that ensures that a beach satisfies stringent ecological, administrative, and safety criteria (https://blueflag.global/ (accessed on 8 May 2022)). An EU Smart Coasts program [31] seeks to maximize the potential of the shoreline on both shores of the Irish Sea in order to safeguard the coastal areas by creating a new real-time water quality prediction system. As part of the service, forecasting water quality based on sampling and analysis of field coastal data connected to an online system provides up-to-date information on water conditions. By complying with EU bathing water regulations that require water samples at certain sites, the technology raises awareness of pollution sources and assists in the upkeep of Blue Flag certified beaches. Another smart coast service may also consider the erosion/accretion of sandy coasts and research to develop a predictive model of submersion to support coastal management in sea-level rise conditions over the next decades [32].

6.2.5. Smart Oceans

Smart monitoring that uses current ocean activities to gather and analyze data is crucial to close the knowledge gap about the open sea and the opportunities it provides [33]. Previously, monitoring technology was restricted by the duration of research vessel expeditions (e.g., battery capacity) and weather conditions, resulting in observations of short-term events or snapshots of longer-term events. Now, smart ocean systems constitute a substantial shift in how scientific research and ocean monitoring are performed. They overcome the limitation of existing technologies by permitting continuous, sub-second observations with a variety of measurement techniques, which are accessible in near real time through the internet to any community. New data solutions are being developed, tested, and implemented, allowing scientists, industry, and users to monitor them from anywhere on the planet.

6.2.6. Smart Marinas

One of the most significant issues for boat captains, particularly during the summer, is the lack of marina reservation online procedures. Most of the time it is not possible to plan a trip by boat securing the position where it will be moored throughout the course of the voyage. A project in Greece [34], financed by the Fiware Accelerator European Commission FrontierCities, arose from this need detected in Greek ports. They created an application for yachters and marina managers to provide e-booking services, navigational, and mooring assistance features. The pilot project, which was carried out at the Patras port yachting area, sought to modernize marina operations by establishing a monitoring system to manage mooring berths, evaluate sea water level, and record meteorological conditions. The project intends to address a need in marina services while also opening a route for smarter communication with the maritime tourism industry, which is conquering new areas for smarter technologies. Smart Marina is the also the name of another EU project [35], whose main mission is to create little guest harbors in the Baltic Sea. The investments in the guest harbors will mostly help toward the rehabilitation of service buildings, new bridges, payment and booking systems, and other types of environmental management in the harbors, all with the goal of creating a favorable environmental profile and a better tourist experience.

6.2.7. Smart Navigation

The SMART-Navigation project is a global initiative that uses the International Maritime Organisation (IMO)'s e-Navigation concept [36], to provide electronic services to Non-SOLAS ships—the ones that are not ruled by the International Convention for the Safety of Life at Sea—such as fishing boats and coastal vessels. The pilot project is focused on Korean maritime traffic providing advanced services, such as:

- Sea traffic management, resulting optimization in vessels traffic flow.

- Knowledge of the maritime domain, which allows vessels to anticipate potentially hazardous circumstances.
- Proactive maritime safety management, with a focus on avoiding identified dangers.
- Remote monitoring, allowing ship systems to be evaluated.
- Telematics service, which provides navigational safety information in a streamlined manner.

Maritime services called Maritime Service Portfolios (MSPs) have been identified by IMO providing operational and technical services for e-Navigation, and in the following table (Table 1) they are mapped to the six (6) services implemented by the SMART-Navigation project.

Table 1. Maritime Service Portfolios (MSPs) linked to SMART-Navigation services.

Code	MSP Service	SMART-Navigation Services
MSP1 MSP2 MSP3 MSP4	VTS Information Service (IS) Navigational Assistance Service (NAS) Traffic Organization Service (TOS) Local Port Service	SV1 → Navigation Monitoring & Assistance Service (NAMAS) SV2 → Ship-borne System Monitoring Service (SBSMS) SV3 → Safe & Optimal Route Planning Service (SORPS)
MSP5	Maritime Safety Information Service (MSI)	SV6 → Maritime Environment and Safety Information Service (MESIS)
MSP6 MSP7	Pilotage service Tug Service	SV5 → Pilot & Tugs Assistance Service (PITAS)
MSP8	Vessel Shore Reporting	
MSP9	Telemedical Assistance Service (TMAS)	
MSP10	Maritime Assistance Service (MAS)	SV1 → Navigation Monitoring & Assistance Service (NAMAS) SV2 → Ship-borne System Monitoring Service (SBSMS)
MSP11	Nautical Chart Service	
MSP12	Nautical Publications Service	
MSP13	Ice Navigation Service	
MSP14 MSP15	Meteorological Information Service Real-time Hydrographic and Environmental Information Service	SV6 → Maritime Environment and Safety Information Service (MESIS)
MSP16	Search and Rescue Service	SV1 → Navigation Monitoring & Assistance Service (NAMAS) SV2 → Ship-borne System Monitoring Service (SBSMS)

7. Implementing the Marine Cadastre

7.1. MSP Implementation Guide

According to UNESCO's report [37], marine spatial planning is "*a process that enables integrated, forward looking, and consistent decision making on the human uses of the sea*". It is a method of providing an ecosystem-based approach for controlling human pressures to the marine environment, similar to land use planning in terrestrial ecosystems. UNESCO has proposed a 10-step guide to accomplish a marine spatial planning endeavor, as shown in Figure 9.

Several of the MSP steps are related to analysis of marine data and their visualization in order to make the necessary decisions. Therefore, the need for establishing an MSDI as a parallel process is of vital importance for the MSP process. Using UNESCO's guide paradigm, the main relationships are highlighted below:

- In step 3, related to pre-planning and establishing planning limitations to organize the process.
- In step 5, for defining and analyzing existing conditions by mapping areas of human activities as well as important ecological areas and identifying spatial conflicts.
- In step 6, which is about defining and analyzing future conditions by mapping new demands for marine space and identifying alternative spatial scenarios.

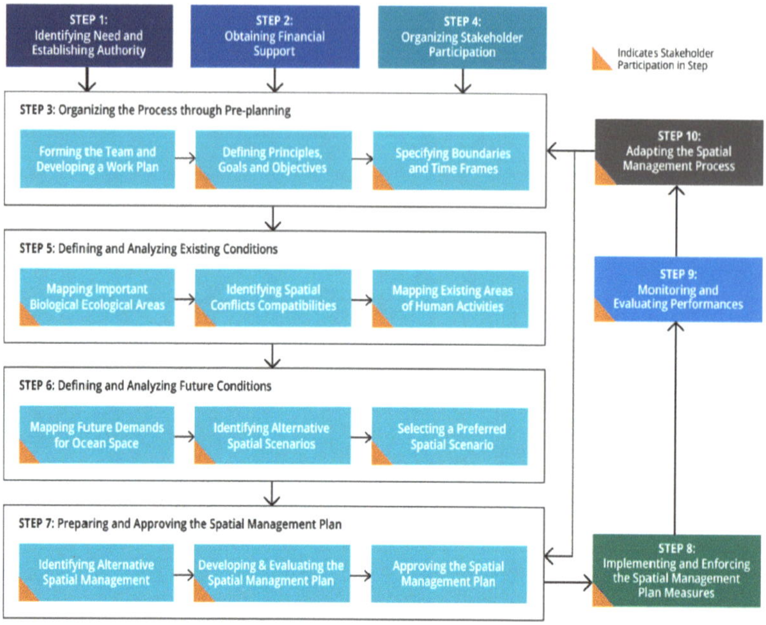

Figure 9. MSP step-by-step guide. Reprinted with permission from Ref. [37], Copyright 2006 UNESCO.

7.1.1. Multipurpose Marine Cadastre

MSP elevates the stakes and broadens the individual states' duties to provide resource capacity management across sea and land interaction. The features of the maritime environment must be recognized and combined into a management system to develop a marine administration system that meets the spatial marine criteria [11]. The marine cadastre is defined [38] as *"a system to enable the boundaries of maritime rights and interests to be recorded, spatially managed and physically defined in relationship to the boundaries of other neighboring or underlying rights and interests."*

Marine cadastre (MC) research articles recognize the three-dimensional (3D) character of marine ecosystems and emphasize the cadastre's need to operate as a multifunctional instrument. The multipurpose maritime cadastre (MMC), an expanded term, has been deployed in the United States [39] as an integrated marine information system that offers jurisdictional, legal, physical, ecological, and human usage data in a shared geographic information system (GIS). It is essentially a data viewer that gives the baseline information required for coastal and marine spatial planning initiatives, notably those involving determining the optimal placement for renewable energy projects. The MMC can also be used during the permit review process. Users can readily view relevant jurisdictional boundaries, restricted areas, laws, sensitive habitat places, and other recorded features by selecting the marine region of interest.

7.1.2. MSDI Establishment

The foundation of a marine spatial data infrastructure (MSDI) in Greece [40] may be closely linked to the realization of the marine cadastre. According to the International Hydrographic Organization [41], *"MSDI is the component of an SDI that encompasses marine geographic and business information in its widest sense. This would typically include seabed topography, geology, marine infrastructure, resource utilisation, administrative and legal boundaries, areas of conservation, marine habitats and oceanography"*. MSDI facilitates the discovery, access, management, distribution, reuse, and preservation of hydrospatial data [42]. In the same way, marine cadastre is recognized as the foundation layer of an MSDI, comprising important marine boundary data as well as associated rights and duties that are continually updated [43,44] (Figure 10). It is important to note that Geographic Information Systems (GIS) provide marine cadastre managers advanced tools for accessing and compiling charts, leveraging information stored in databases, and automating relevant processes [45].

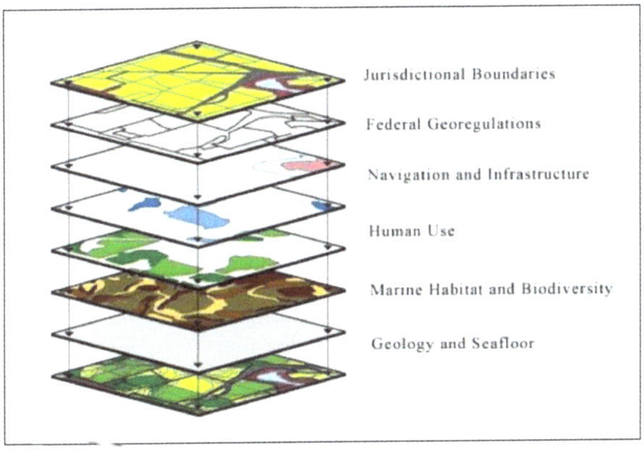

Figure 10. Data layers in an MSDI. Reprinted from Ref. [44]. Copyright 2010, Fowler et al.

The function of the marine cadastre as a data layer in a marine SDI has been discussed since the international workshop on regulating the marine environment held in Malaysia in 2004 [46]. The workshop proposed that, as an analogue to a "land administration system," the name "marine administration system (MAS)" to be used for the *"administration of rights, restrictions and responsibilities in the marine environment with the spatial dimension facilitated by the Marine SDI"*.

7.1.3. MSDI Enablers

The availability of spatial information content to users is the most significant component of MSDI, as it is of little value without it. The data shall be presented within the context of a consistent coordinate reference system. At the heart of this content is reference information, which refers to the most commonly used datasets, themes, or spatial data layers, which together form a digital base chart that can be portrayed and searched. According to the International Hydrographic Organization (IHO) [41], the following MSDI capabilities listed below are regarded as important building elements that serve as the foundation for data collection, administration, modification, and dissemination:

- **Standards**: The ISO 19100 series of international geographic standards, as well as the Open Geospatial Consortium (OGC) standards, are critical to building a strong SDI architecture, particularly in the areas of data content modeling, data transfer, and web services.

- **Technology**: The availability of technological infrastructure supports the delivery of data and services that enable data reading, exchange, conversion, and dissemination to enhance informational goods. As the infrastructure improves, the SDI may be able to function not just in multiple geodetic schemes, but also to convert data to create informational content in various projections.

- **Metadata**: They are "data about data" that indicate the qualities of a dataset (for instance, content, value, and restrictions) and are generally kept in a metadata management system to enable information extraction capabilities. It is essential for locating data and information as well as knowing how the data may be utilized. A web site is the most common way for individuals to quickly and easily search for content using its metadata.

- **Universal Hydrographic Data Model**: IHO's S-100 series of standards provides the data framework for the production of relevant digital products and services needed by marine communities [47]. S-100-based product specifications are currently available for a variety of marine data services, such as S-121 for Maritime Limits and Boundaries (MLBs) and S-122 for Marine Protected Areas (MPAs).

7.1.4. S-121 Maritime Limits and Boundaries

The Maritime Limits and Boundaries standard (S-121) is designed for the encoding and sharing of digital maritime boundary information, such as maritime limits, zones, and boundaries as defined by the United Nations Convention on the Law of the Sea (UNCLOS) [13,48]. International boundaries, coastlines, internal waterways, territorial waters, contiguous and exclusive economic zones, and the continental shelf are all covered under S-121. Additional entities that nations should declare (for example, joint development regions) as a result of bilateral treaties should be defined for use in a marine cadastre system [49]. Each real-world feature is represented as an object in S-121, with properties represented as attributes (both geographic and thematic) and associations that give context for the feature.

The four major components of each MLB object are:

1. The party component which defines the different actors and their role associated with an object.
2. The geospatial component that specifies the object's location and type.
3. The legal component, which supports the definition of the related jurisdictions and rights in relation to the object.
4. Administrative or geographical sources such as treaties, legal papers, and maps.

The MLB model is sufficiently extensive whether implemented in a geospatial system or as part of an MSDI to facilitate the production of different products and services (http://www.s-121.com/ (accessed on 8 May 2022)). These could include the deposit of national maritime boundary claims or the compilation of maritime boundary objects for inclusion in S-57 and S-101 Electronic Navigational Charts. As stated in this article [11], S-121 focuses on the legal description of marine entities. Various definitions of legal rights, as well as associated constraints and obligations, can be created for different parties, even if these parties have potentially competing claims.

7.1.5. Land Administration Alignment

S-121 makes use of ISO-19152, which establishes a reference land administration domain model (LADM). ISO-19152 allows for the legal definition of associated rights, restrictions, and duties, as well as proper reference via sourcing and versioning, and this feature connects the standard with legal traceability procedures. The adoption of the ISO-19152 standard leverages the large community involvement in land administration, which has many similarities with the management of maritime boundaries and limits. The application of an LADM lays the groundwork for extending S-121 into the management of all other regulated boundaries, such as marine reserves and fisheries. The compatibility with the land domain model promotes uniform administration of the littoral zone for states that use S-121 for maritime areas and ISO-19152 for land jurisdiction.

A marine rights data model is described in [49,50] and provides a common means of capturing the rules that assist the assignment, demarcation, documentation, evaluation, and selection of marine property rights, awarded interests, resources available, and their 3D spatial extent (Figure 11). The section aims to represent regulatory, organizational, and environmental elements that are commonly associated with marine parcels, and the relevant design is based on the rationale that rights, obligations, and constraints in marine spaces relate to explicitly 3D/4D space, i.e., the sea surface, water column, seabed, and seafloor subsurface.

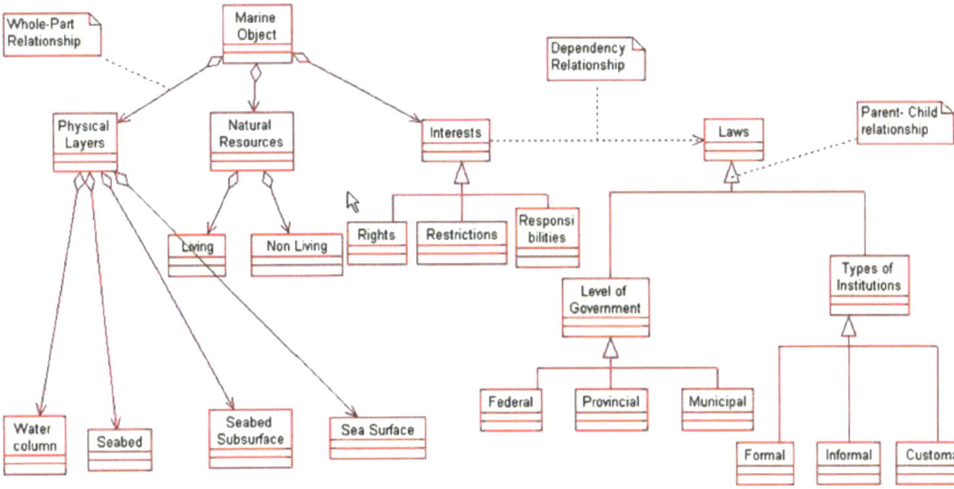

Figure 11. A marine parcel data model. Reprinted from Ref. [49]. Copyright 2014, Ng'ang'a et al.

7.1.6. S-122 Marine Protected Areas (MPAs)

The Marine Protected Areas product specification (S-122) is designed to encapsulate MPA information for use in MSDIs [50]. MPAs are seas, oceans, rivers, or lakes that have been designated as protected areas. They may include regions of intertidal or subtidal topography, as well as their underlying water and related flora, animal, historical, and cultural aspects, that have been reserved by law or other effective ways to safeguard a portion or the entirety of the enclosed ecosystem. MPAs could be designed to protect unique fish species, scarce habitat areas, or entire ecosystems. MPAs can range from basic declarations to protect a single resource to highly regulated areas.

The extent to which environmental standards affect maritime operations varies depending on whether MPAs are located in territorial waters, exclusive economic zones, or the high seas. The majority of MPAs are in the territorial waters of coastal nations, where enforcement is possible. MPAs can also be established in a state's exclusive economic zone and even within international waters. Italy, France, and Monaco, for example, have collaborated to build the Pelagos Sanctuary for Mediterranean Marine Mammals in the Ligurian Sea. This refuge encompasses both international and domestic seas.

7.2. MSDI Cloud Infrastructure

The public-based company Information Society S.A., in the context of the modernization of the public sector ICT infrastructure in Greece, designed and implemented the project for government cloud computing named Government Cloud, or in short, G-Cloud. It offers digital services based on state-of-the-art cloud computing and virtualization infrastructures. G-Cloud intends to share computing resources among government agencies, lowering purchase, operation, and service costs while enhancing flexibility and security, with the purpose of improving services to citizens and organizations. Figure 12 illustrates the three G-Cloud types of services and the stack of the components they include. MSDI could be offered either as platform as a service (PaaS) or software as a service (SaaS) type of G-Cloud, if specific "smart" services are offered to the various stakeholders.

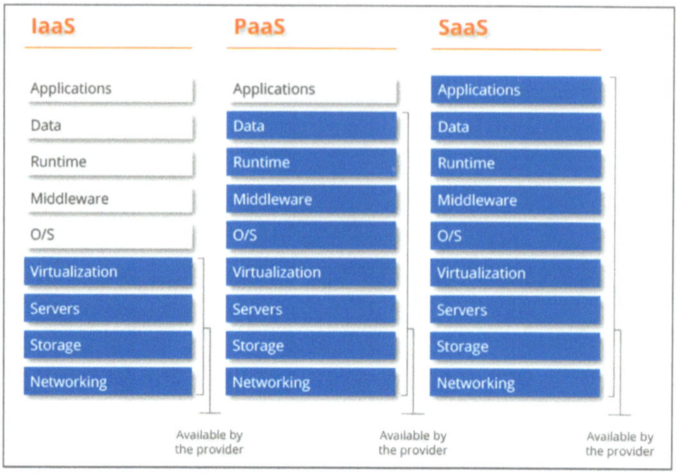

Figure 12. G-Cloud type of services in Greece. Reproduction from https://www.ktpae.gr/erga/government-cloud-g-cloud/ (accessed on 08 May 2022).

8. Discussion

In June 2018, Law 4546/2018 (Government Gazette 101/A/12-06-2018) was incorporated into the Greek legal order the Directive 2014/89/EU, setting out the general framework for the application of MSP for the sustainable development of the maritime economy, development of maritime areas, and sustainable use of resources. A defined framework for the spatial organization of maritime activities can be expected as a result of the MSP and the institutionalization of maritime spatial planning. However, maritime activity is heavily reliant on land, with the anticipation of further activities, which complicates the matter of both obligations and planning in general, as is the case of Greece.

The European Union's deadline on April 2021 for Member States to develop maritime spatial plans passed without the needed steps being completed for Greece [51]. The country demonstrated a lack of readiness on this critical issue in order to meet not only its European obligations, but also the responsibility to protect the Greek seas and to achieve sustainable "blue" development. However, Greece was only beginning to recover from a decade-long economic crisis, that had taken a fourth off its GDP, when the COVID-19 pandemic erupted. Although the pandemic has slowed the country's recovery, it has also provided policymakers with the opportunity to focus on how to bring the economy back on track with more sustainable post-COVID development. The National Recovery and Resilience Plan, called Greece 2.0, is the government's response to tackle similar issues, an extensive package of investments and reforms in major sectors of the economy. The plan includes the development of urban plans and the development of new spatial planning for renewables, industry, tourism, and aquaculture, as well as marine spatial planning in general [52].

According to the Hellenic Federation of Enterprises (SEV) research on MSP [3], marine spatial planning should strike the correct balance between the country's marine ecosystem's sustainability and the large investments that can be made at sea and shore. At a time when the country is progressively returning to the focus of investment, the state must establish the groundwork for an efficient MSP that avoids duplication and conflict with existing marine and coastal policies (local spatial plans, regional spatial plans, special spatial planning and sustainable development frameworks for industry, tourism, renewable energy, and aquaculture). It is also critical to provide enhanced project coherence and coordination in order to minimize duplication and/or contradictory spatial approaches between different levels of planning. Marine spatial planning shall be [3,24]:

- Multi-objective and integrated, embracing all main economic sectors, having economic, social, and environmental targets.

- Strategic and forward-thinking, exploring different methods of achieving a vision.
- Ongoing and adaptable, with a focus on performance assessment and acquiring knowledge upon doing.
- Participatory, building a diverse stakeholders base to guarantee long-term management commitment.
- Ecosystem-based, with an emphasis on long-term environmental resource preservation.
- Geographically focused, with an emphasis on marine zones that people could understand, connect to, and be concerned about.

In order to achieve the aforementioned objectives, the necessary reforms and investments in Greece shall include the establishment of a marine cadastre and an MSDI. Both facilitate MSP and ICZM processes in making marine ecosystem-based management a reality, such as, for instance, in designating marine protected areas [53]. MSDI shall provide timely access to data from public and private organizations in marine-related disciplines such as hydrography, oceanography, meteorology, and maritime economic sectors [42]. Furthermore, within a marine cadastre, being a well-built geographic information system (GIS), shall be recorded the boundaries of maritime rights and interests, spatially managed and physically defined in relationship to the boundaries of other neighboring underlying rights and interests. A marine cadastre shall be a base layer in the MSDI that public authorities shall rely on as the official infrastructure to access and integrate multi-source marine spatial data.

9. Conclusions

Smart marine ecosystem-based planning (SMEP) has been introduced in this study as a framework for more participatory and responsive maritime administration, aiming at safer marine ecosystems, and better fulfilling the needs of the marine stakeholders through "smart" hydrospatial data services. According to the World Wildlife Fund (WWF) [54], identifying trends in marine species and activities, as well as accounting for ecosystem capability and the possibility for recovery from human-caused changes, necessitates long-term data on environmental variables and human activities. The SMEP strategy and the MSDI realization [24] are key success factors towards this direction that could guarantee greater legal certainty and unleash sustainable growth momentum. SMEP is an adaptive strategic framework to incorporate changes, being driven by the environment and the climate change forces, the blue growth economy targets of the respective EU programming periods, and the implications from the geopolitical chess of the East Mediterranean Sea basin. All these factors require deep knowledge and rational decisions for achieving collaboration for the Eastern Mediterranean's long-term sustainable development, that SMEP strategy and MSDI realization aim to provide. Nautical cartography standards such as S-121 and S-122, as well as hydrospatial and maritime data services have been recently defined by IHO and IMO, respectively, to guide the recording, planning, and management processes, where the spatial extent of rights, restrictions, and responsibilities in the marine environment need to be defined.

Author Contributions: Conceptualization, S.C.; methodology, S.C. and A.P.; investigation, S.C.; writing—original draft preparation, S.C.; writing—review and editing, A.P. and B.N.; visualization, S.C.; supervision, B.N. All authors have read and agreed to the published version of the manuscript.

Funding: This research received no external funding.

Conflicts of Interest: The authors declare no conflict of interest.

References

1. United States Defense Mapping Agency. Coastal Length Data Are Based on the World Vector Shoreline. Figures Were Calculated by Pruett, L. and Cimino, J., Unpublished Data, Global Maritime Boundaries Database (GMBD), Veridian—MRJ Technology Solutions, (Fairfax, Virginia, January 2000). 1989. Available online: https://web.archive.org/web/20120419075053/http://earthtrends.wri.org/text/coastal-marine/variable-61.html (accessed on 8 May 2022).
2. Nakos, B.; Filippakopoulou, V.; Stamou, L. The impact of line generalization on the length of Greek coastlines. In Proceedings of the 7th National Cartographic Conference, Hellenic Cartographic Society, Mytilini, Greece, 23–26 October 2002; pp. 279–290. (In Greek)
3. Hellenic Federation of Enterprises (SEV). Special Expertise Report on Adaptation of Entrepreneurship Policies and Processes from the Establishment of a Specific Institutional Framework for Marine Spatial Planning, Athens, October 2018. Available online: https://www.sev.org.gr/Uploads/Documents/52131/MELETI_final.pdf (accessed on 8 May 2022). (In Greek)
4. Lawrence, D.; Kenchington, R.; Woodley, S. *The Great Barrier Reef*; Melbourne University Press: Melbourne, Australia, 2002.
5. Greece Urged to Protect Hellenic Trench from Seismic Blasts. Available online: https://www.nrdc.org/experts/francine-kershaw/greece-urged-protect-hellenic-trench-seismic-blasts (accessed on 8 May 2022).
6. Frantzis, A.; Leaper, R.; Alexiadou, P.; Prospathopoulos, A.; Lekkas, D. Shipping routes through core habitat of endangered sperm whales along the Hellenic Trench, Greece: Can we reduce collision risks? *PLoS ONE* **2019**, *14*, e0212016. [CrossRef] [PubMed]
7. Ionian Archipelago, Important Marine Mammal Area. Available online: https://www.marinemammalhabitat.org/portfolio-item/ionian-archipelago/ (accessed on 8 May 2022).
8. Bradshaw, C.J.A.; Greenhill, L.; Yates, K.L. The future of marine spatial planning. In *Offshore Energy and Marine Spatial Planning*; Routledge: Abingdon, UK, 2018; pp. 284–293. [CrossRef]
9. European Commission. *Communication from the Commission to the Council and the European Parliament on Integrated Coastal Zone Management: A strategy for Europe*; COM(2000) 547; Publications Office of the European Union: Brussels, Belgium, 2000.
10. Kristina, V.; Pille, M. Assessment of Application of Baltic Sea Common Regional Maritime Spatial Planning Framework. Pan Baltic Scope Project Report Commissioned by VASAB. 2019. Available online: https://vasab.org/documents/msp-and-iczm/ (accessed on 8 May 2022).
11. HELCOM-VASAB MSP Working Group. Baltic Sea Broad-Scale Maritime Spatial Planning Principles. Available online: https://helcom.fi/media/documents/HELCOM-VASAB-MSP-Principles.pdf (accessed on 11 July 2021).
12. Ventura, A.M.F. *Environmental Jurisdiction in the Law of the Sea*; Springer International Publishing: Berlin/Heidelberg, Germany, 2020. [CrossRef]
13. Kastrisios, C.; Tsoulos, L. Maritime zones delimitation—Problems and solutions. In Proceedings of the International Cartographic Association Conference, Washington, DC, USA, 2–7 July 2017; Volume 1, pp. 1–7. [CrossRef]
14. Athanasiou, K.; Sutherland, M.; Kastrisios, C.; Tsoulos, L.; Griffith-Charles, C.; Davis, D.; Dimopoulou, E. Toward the Development of a Marine Administration System Based on International Standards. *ISPRS Int. J. Geo-Inf.* **2017**, *6*, 194. [CrossRef]
15. European Commission. *An Integrated Maritime Policy for the European Union*; Commission Communication COM(2007) 575; Publications Office of the European Union: Brussels, Belgium, 2007.
16. European Commission. *Blue Growth, Opportunities for Marine and Maritime Sustainable Growth*; Commission Communication COM(2010) 494; Publications Office of the European Union: Brussels, Belgium, 2012.
17. European Commission. *Europe 2020. A Strategy for Smart, Sustainable and Inclusive Growth*; COM (2010) 2020 Final; European Commission: Brussels, Belgium, 3 March 2010.
18. European Commission. *Directorate-General for Maritime Affairs and Fisheries, Marine Knowledge 2020: From Seabed Mapping to Ocean Forecasting: Green Paper*; Publications Office of the European Union: Brussels, Belgium, 2012; Available online: https://data.europa.eu/doi/10.2771/4154 (accessed on 8 May 2022).
19. Shepherd, I. European efforts to make marine data more accessible. *Ethics Sci. Environ. Politics* **2018**, *18*, 75–81. [CrossRef]
20. European Parliament and Council Directive 2014/89/EU of the European Parliament and of the Council of 23 July 2014 Establishing a Framework for Maritime Spatial Planning. 2014. Available online: http://data.europa.eu/eli/dir/2014/89/oj (accessed on 8 May 2022).
21. Incorporating the MSP EU Directive in the Greek Law (Greek Text). Available online: https://www.alikakou.gr/gr/el/articles/thalassios-xorotaksikos-sxediasmos-i-ensomatosi-tis-odigias-2014-89ee-me-to-to-n-45462018 (accessed on 8 May 2022).
22. Ramieri, E.; Bocci, M.; Markovic, M. Linking Integrated Coastal Zone Management to Maritime Spatial Planning: The Mediterranean Experience. In *Maritime Spatial Planning*; Springer International Publishing: Berlin/Heidelberg, Germany, 2019; pp. 271–294. [CrossRef]
23. Papageorgiou, M. Marine Spatial Planning and Marine Uses: Conceptual and Theoretical Issues. *J. Aeixoros* **2016**, *23*, 41–63.
24. Moraitakis, G. Smart Marine Ecosystem, Safety4Sea Conference, Presentation of Wärtsilä Greece. 2019. Available online: https://safety4sea.com/cm-smart-marine-ecosystem/ (accessed on 8 May 2022).
25. IPCC. Summary for Policymakers. In *Climate Change 2021: The Physical Science Basis*; Contribution of Working Group I to the Sixth Assessment Report of the Intergovernmental Panel on Climate Change; IPCC: Geneva, Switzerland, 2021; Available online: https://www.ipcc.ch/report/ar6/wg1/ (accessed on 8 May 2022).
26. Contarinis, S.; Pallikaris, A.; Nakos, B. The Value of Marine Spatial Open Data Infrastructures—Potentials of IHO S-100 Standard to Become the Universal Marine Data Model. *J. Mar. Sci. Eng.* **2020**, *8*, 564. [CrossRef]

27. Molavi, A.; Lim, G.J.; Race, B. A framework for building a smart port and smart port index. *Int. J. Sustain. Transp.* **2019**, *14*, 686–700. [CrossRef]
28. Katikas, L.; Kontos, T.D. Suitability Assessment for Offshore Wind Farm Siting Using Exclusion Criteria and Spatial Economic Models: The Case of North and Central Aegean Sea. In Proceedings of the Environmental Science, Policy & Management, Lesvos Island, Greece, 1–3 June 2018. [CrossRef]
29. Yang, Y.; Yang, X.; Tan, J.; Chen, L.; Zheng, M.; Wang, S. The Scheme Design of Smart Offshore Wind Farm. In Proceedings of the 2019 Prognostics and System Health Management Conference (PHM-Qingdao), Qingdao, China, 25–27 October 2019; IEEE: Piscataway, NJ, USA, 2019; pp. 1–6.
30. Dzulqornain, M.I.; Rasyid MU, H.A.; Sukaridhoto, S. Design and Development of Smart Aquaculture System Based on IFTTT Model and Cloud Integration. *MATEC Web Conf.* **2018**, *164*, 01030. [CrossRef]
31. Smart Coasts: Measuring Water Quality at Welsh and Irish Beaches. Available online: https://ec.europa.eu/regional_policy/it/projects/united-kingdom/smart-coasts-measuring-water-quality-at-welsh-and-irish-beaches (accessed on 8 May 2022).
32. Scardino, G.; Sabatier, F.; Scicchitano, G.; Piscitelli, A.; Milella, M.; Vecchio, A.; Anzidei, M.; Mastronuzzi, G. Sea-Level Rise and Shoreline Changes Along an Open Sandy Coast: Case Study of Gulf of Taranto, Italy. *Water* **2020**, *12*, 1414. [CrossRef]
33. Smart Ocean. Available online: https://goexplorer.org/smart-ocean/ (accessed on 8 May 2022).
34. Platform for the Yachting Industry, Interconnecting Marinas and Yachters. Available online: https://www.fiware.org/success-stories/sammy/ (accessed on 8 May 2022).
35. Smart Marina Project. Available online: https://www.smartmarina.eu (accessed on 8 May 2022).
36. SMART-Navigation Project. Available online: https://www.iala-aism.org/technical/e-nav-testbeds/smart-navigation-project/ (accessed on 8 May 2022).
37. Ehler, C.; Douvere, F. *Visions for a Sea Change: Report of the First International Workshop on Marine Spatial Planning*; Intergovernmental Oceanographic Commission and the Man and the Biosphere Programme UNESCO Headquarters: Paris, France, 2006. [CrossRef]
38. Principles for a Seabed Cadastre. OSG Technical Report 9, Office of the Surveyor-General. New Zealand, 1999. Available online: https://docplayer.net/159576845-Principles-for-a-seabed-cadastre-osg-technical-report-9-office-of-the-surveyor-general.html (accessed on 8 May 2022).
39. Multipurpose Marine Cadastre. NOAA, Bureau of Ocean Energy Management. 2012. Available online: http://dusk.geo.orst.edu/ICAN_EEA/Members/NOAA_MMC.pdf (accessed on 8 May 2022).
40. Contarinis, S.; Nakos, B.; Pallikaris, A. Marine Spatial Data Infrastructure in Greece—Challenges and Opportunities. *Nausivios Chora* **2018**, *7*, E3–E30. Available online: http://nausivios.hna.gr/docs/2018E1.pdf (accessed on 8 May 2022).
41. International Hydrographic Organisation (IHO). *Spatial Data Infrastructures "The Marine Dimension"*; Guidance for Hydrographic Offices, IHO Publication C-17—Edition 2.0.0; International Hydrographic Bureau: Monte-Carlo, Monaco, 2017.
42. Contarinis, S.; Kastrisios, C. Marine Spatial Data Infrastructure. In *The Geographic Information Science & Technology Body of Knowledge (1st Quarter 2022 Edition)*; Wilson, J.P., Ed.; University Consortium for Geographic Information Science (UCGIS): Ithaca, NY, USA, 2022. [CrossRef]
43. FIG. *Administering Marine Spaces: International Issues*; FIG Report; The International Federation of Surveyors (FIG): Frederiksberg, Denmark, 2006; ISBN 87-90907-55-8.
44. Fowler, C.; Smith, B.F.; Stein, D. *Building a Marine Spatial Data Infrastructure to Support Marine Spatial Planning in US Waters*; National Oceanic and Atmospheric Administration (NOAA), Coastal Services Center: Charleston, SC, USA, 2010; 6p.
45. Sun, J.; Mi, S.; Olsson, P.-O.; Paulsson, J.; Harrie, L. Utilizing BIM and GIS for Representation and Visualization of 3D Cadastre. *ISPRS Int. J. Geo-Inf.* **2019**, *8*, 503. [CrossRef]
46. Balla, E.; Wouters, R. Marine Cadastre in Europe: State of play (NR 355). In *World Bank Conference on Land and Poverty*; The World Bank: Washington, DC, USA, 2017.
47. IHO. S-100 Universal Hydrographic Data Model (Edition 4.0.0, December 2018). Available online: https://iho.int/uploads/user/pubs/standards/s-100/S-100_Ed%204.0.0_Clean_17122018.pdf (accessed on 8 May 2022).
48. IHO. S-121 Maritime Limits and Boundaries Product Specification v1.0.0. 2019. Available online: https://registry.iho.int/productspec/view.do?idx=177&product_ID=S-121 (accessed on 8 May 2022).
49. Ng'ang'a, S.M.; Sutherland, M.; Cockburn, S.; Nichols, S. Toward a 3D marine Cadastre in support of good ocean governance: A review of the technical framework requirements. *Comput. Environ. Urban Syst. J.* **2004**, *28*, 443–470. [CrossRef]
50. IHO. S-122 Marine Protected Areas Product Specification v1.0.0. 2019. Available online: https://registry.iho.int/productspec/view.do?idx=73&product_ID=S-122 (accessed on 8 May 2022).
51. WWF. Two-Thirds of EU Countries Fail to Submit Plans for Sustainable Management of Their Seas on Time. 2021. Available online: https://www.wwf.eu/wwf_news/media_centre/?uNewsID=2717941 (accessed on 8 May 2022).
52. Hellenic Republic. Greece 2.0 National Recovery and Resilience Plan. 2021. Available online: https://primeminister.gr/wp-content/uploads/2021/03/Greece-2_0-April-2021.pdf (accessed on 8 May 2022).
53. Douvere, F. The importance of marine spatial planning in advancing ecosystem-based sea use management. *Mar. Policy* **2008**, *32*, 762–771. [CrossRef]
54. WWF. Position Paper: Achieving Ecosystem-Based Marine Spatial Plans. 2020. Available online: https://wwfeu.awsassets.panda.org/downloads/wwf_position_paper_ecosystem_based_approach_in_msp___feb2020.pdf (accessed on 8 May 2022).

geomatics

MDPI

Article

Classification of Coastal Benthic Substrates Using Supervised and Unsupervised Machine Learning Models on North Shore of the St. Lawrence Maritime Estuary (Canada)

Guillaume Labbé-Morissette *,†, Théau Leclercq *,†, Patrick Charron-Morneau, Dominic Gonthier, Dany Doiron, Mohamed-Ali Chouaer and Dominic Ndeh Munang

Interdisciplinary Centre for the Development of Ocean Mapping (CIDCO), 115 Rue St Germain O local 1, Rimouski, QC G5L 4B6, Canada; patrick.cmorneau@cidco.ca (P.C.-M.); dominic.gonthier@cidco.ca (D.G.); dany.doiron@cidco.ca (D.D.); mohamed-ali.chouaer@cidco.ca (M.-A.C.); mdndeh@natobservations.com (D.N.M.)

* Correspondence: guillaume.morissette@cidco.ca (G.L.-M.); theau.leclercq@uqar.ca (T.L.); Tel.: +1-418-725-1732 (ext. 1722) (G.L.-M.); +1-418-896-0919 (T.L.)

† These authors contributed equally to this work.

Abstract: Classification of benthic substrates is a core necessity in many scientific fields like biology, ecology, or geology, with applications branching out to a variety of industries, from fisheries to oil and gas. In the first part, a comparative analysis of supervised learning algorithms has been conducted using geomorphometric features to generate benthic substrate maps of the coastal regions of the North Shore of Quebec in order to establish a quantitative assessment of performance to serve as a benchmark. In the second part, a new method using Gaussian mixture models is showcased on the same dataset. Finally, a side-by-side comparison of both methods is featured to provide a qualitative assessment of the new algorithm's ability to match human intuition.

Keywords: benthic habitat mapping; benthic habitat classification; supervised machine learning; unsupervised machine learning; artificial intelligence

Citation: Labbé-Morissette, G.; Leclercq, T.; Charron-Morneau, P.; Gonthier, D.; Doiron, D.; Chouaer, M.-A.; Munang, D.N. Classification of Coastal Benthic Substrates Using Supervised and Unsupervised Machine Learning Models on North Shore of the St. Lawrence Maritime Estuary (Canada). *Geomatics* **2024**, *4*, 237–252. https://doi.org/10.3390/geomatics4030013

Academic Editor: Christophe Claramunt

Received: 21 March 2024
Revised: 29 May 2024
Accepted: 26 June 2024
Published: 30 June 2024

1. Introduction

According to the Organization for Economic Cooperation and Development (OECD), the ocean economy will reach three trillion dollars by 2030. In contrast, almost 75% of the world's oceans are not mapped to modern standards [1]. In the spirit of increasing our knowledge of the oceans, benthic habitat mapping has become a necessity with very high stakes, with many countries, notably Canada, engaging in massive benthic mapping campaigns [2]. While the world requires more data-driven decision-making to ensure proper sustainable stewardship of natural resources on one end, the efficiency requirements of commercial and industrial ventures have never been higher. As such, new technologies are required to adequately map out the benthic zones efficiently. Since diver-based mapping, remotely operated vehicles, and other in situ methods would prove themselves to be too costly to map out the entirety of the world's oceans, remote sensing methods have become a staple of the habitat mapping community. Of particular economic interest are geomorphometric methods which can be generated from a wide array of acoustic and optical remote sensing sources that generate point cloud data in three dimensions. As such, we provide a comparative analysis of several machine learning algorithms applied to point cloud data generated in the context of multiple multibeam echosounder (MBES) surveys conducted on the North Shore of the St. Lawrence maritime estuary.

2. Background

Traditionally, benthic substrate identification began with direct observation methods. These include diver surveys and sediment sampling [3,4]. Divers qualitatively assess the substrate type, flora, and fauna, providing detailed but spatially limited data. Sediment grabs and core samples allow for precise textural and compositional analysis of substrates in a laboratory setting, offering insights into sediment characteristics and distributions. However, while they remain a popular option, their limited spatial coverage makes them inadequate to classify large areas.

To extend the spatial coverage of these techniques, acoustic methods have become fundamental in large-scale substrate mapping. These include single-beam and multibeam sonar systems, which emit sound waves and analyze the returned signals to infer substrate characteristics based on acoustic properties such as the return signal's strength (backscatter) [5], and timeseries [6]. Side-scan sonar, particularly, produces detailed images of the seabed, enabling the identification of substrate types and benthic features across extensive areas [7].

Advancements in optical technologies have led to the increased use of underwater photography and videography. These methods provide high-resolution, direct visual accounts of the benthic environment. A particular application of optical advancements can be found in remote sensing satellites. While less detailed than direct methods and limited to relatively shallow waters, remote sensing has proven itself to be invaluable in mapping large, remote marine areas [8]. These methods classify substrate types based on their reflectance intensities at various wavelengths using either discrete bands in the case of multispectral imaging [9], or continuous spectrum in the case of hyperspectral imaging [10].

All imaging methods, both acoustic and optical, can be further exploited using automated image analysis algorithms. These include classical computer vision algorithms such as co-occurence matrixes [11], traditional machine learning algorithms such as random forests [12], object-based image analysis [13], and deep-learning methods [14,15].

In addition to image-based methods, several remote sensing technologies such as multibeam echosounders, LiDAR and satellite-derived bathymetry are able to provide point clouds in three dimensions. These point clouds can be processed into digital terrain models (DTM) onto which numerical methods such as geomorphometry can be applied [16]. While originally developed for land-based models, these techniques have been successfully transferred to the hydrographic world [17], and are now part of the established literature on the subject [18].

3. Area of Interest

The St. Lawrence River and estuary, a critical hydrological system in North America, holds substantial importance from a scientific standpoint due to its strategic location as the entry point to the Great Lakes ecosystem. Spanning approximately 3058 km, it connects the Great Lakes to the Atlantic Ocean, serving as a vital conduit for water flow, nutrient cycling, and sediment transport. It is also the main maritime transport route for the Great Lakes ecosystem, which connects large areas of economic interest such as Ontario, Quebec, Illinois, Michigan, Ohio, Wisconsin, Indiana, Minnesota, New York and Pennsylvania.

The St. Lawrence River supports a large array of aquatic and terrestrial ecosystems and provides unique habitats for a large number of species, including commercially significant populations such as the Atlantic salmon (Salmo salar) and the Snow Crab (Chionoecetes opilio). Its wetlands are home to a wide variety of migratory birds, offering shelter, breeding, and feeding areas.

Hydrologically, the St. Lawrence River is a critical component of the Great Lakes–St. Lawrence Basin. As one of the largest freshwater systems in the world, it regulates the water levels of the Great Lakes and influences both upstream and downstream hydrological conditions. As such, it plays a critical role in flood control, water supply, and hydroelectric power generation.

Additionally, the St. Lawrence River is essential to regional and global chemical cycles. It acts as an essential pathway for carbon, nitrogen, and phosphorus transport from terrestrial to marine environments, a path that is critical to the health of aquatic food chains and ecosystems. The river's sediments are studied for their role in sequestering contaminants and their potential as pollution sources under changing environmental and climatic conditions.

Finally, the St. Lawrence River and its estuary are an extensive laboratory for the study of the impacts of climate change. Alterations in temperature, meteorological patterns, and ice cover affect the river's hydrology, chemistry, and ecology. Predicting changes in flow regimes, increased frequency of extreme weather events, and shifts in species distributions are essential for developing adaptive management strategies.

4. Materials and Methods

The two proposed methods rely on classifying each sounding based on geomorphometric features computed on its spatial neighborhood. For the first method, we use supervised learning with ground-truthing data from Fisheries and Oceans Canada (DFO). We obtain a training set of soundings that can be used to train various supervised models, which can subsequently be used to classify out-of-band data. We use the bulk 80% to train the models, and the remaining 20% to assess the quality of the models. For the unsupervised model, ground-truth data were not necessary since the model was trained and computed with a Gaussian Mixture Method on all data.

4.1. Data Sources

4.1.1. Multibeam Echosounder Data

Multibeam echosounders (MBES) work by emitting a fan-shaped array of acoustic beams from a transducer mounted on a vessel. The transducer sends out multiple beams simultaneously, typically a few dozen to several hundred, covering a wide swath of the seafloor perpendicular to the vessel's path. As the sound waves travel through the water, they eventually encounter the seafloor or other underwater objects. Upon hitting these surfaces, the sound waves are reflected back towards the transducer as echoes, which are then sensed back by the transducer. The time taken for each beam to travel to the seafloor and back is used along with the speed of sound in water to compute the distance from the transducer to the seafloor. Measurements from a sound velocity probe are used to adequately model the speed of sound in water, which can vary based on factors like temperature, salinity, and depth. By continuously transmitting sound pulses and moving the ship, the MBES collects depth information across a wide section of the seafloor. The width of this section is typically several times the depth, allowing for efficient and comprehensive mapping.

To acquire depth measurements with high positional accuracy, the MBES requires a high-accuracy precision source. To this end, a global navigation satellite system (GNSS) provides a steady stream of position measurements centered at the ship's antenna's phase center. To include the ship's alignment with regard to the seafloor and apply corrections based on the ship's movement in three dimensions, an inertial navigation system (INS) is used to measure the roll, pitch and heading angles of the vessel, along with the linear and angular accelerations of the vessel with regards to each axis. These accelerations are integrated twice by the INS, and fused with the GNSS readings using a Kalman filter to provide a robust and accurate estimate of the ship's position in real time.

This position stream's accuracy can be further enhanced using multiple correction methods. These corrections include errors caused by atmospheric conditions, satellite clock errors, and many more. Real-time kinematics (RTK) uses a continuous stream of signal corrections sent to the GNSS receiver to achieve centimeter-level accuracy in real time. While very practical due to its ability to provide a measurement in the field, this method is vulnerable to signal interruption and interference that can occur during the survey. As a more resilient method, post-processed kinematics (PPK) using recorded base-station data after the survey can be used to provide a more accurate trajectory estimate.

In the case of the surveys presented here, high-density MBES data have been gathered over several coastal regions of interest on the North Shore (Figure 1). The study areas were surveyed using CIDCO's hydrographic vessel, the FJ Saucier. The vessel's hydrographic system is comprised of an IxBlue Hydrins inertial navigation system (INS), a Septentrio AsteRx-U GNSS, and a Reson SeaBat 7125 MBES. The system's static calibration and offset measurements were conducted using a total station. The system's dynamic calibration and boresight angles were measured using standard IHO patch-test methodology.

The positioning and attitude data were fused and corrected using real-time kinematics (RTK) and post-processed kinematics (PPK) whenever base station data were available. The bathymetry was computed using the CARIS HIPS and SIPS version 11.4 software suite. The raw bathymetry was smoothed using the CUBE algorithm [19], and then decimated and interpolated to a 1 meter by 1 meter digital terrain model (DTM).

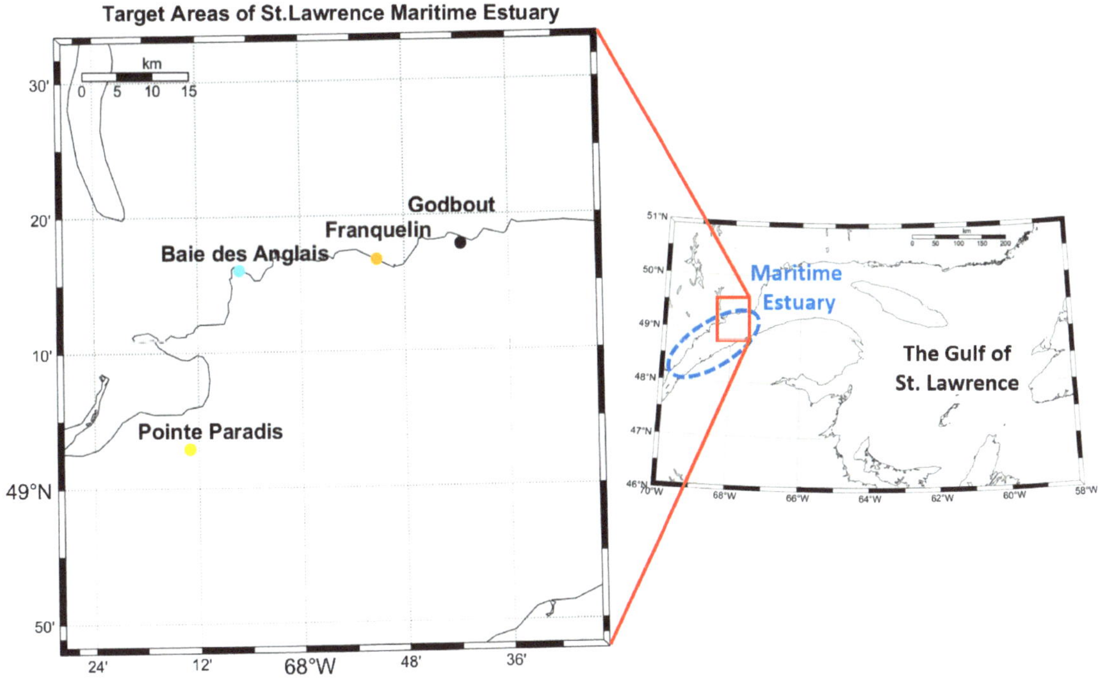

Figure 1. Study zones.

4.1.2. Ground-Truth Data

Fisheries and Oceans Canada (DFO) has developed a dictionary of underwater habitats, with data acquired at 905 coastal ground-truth stations (Figure 2). The imagery at these stations has been acquired using drop camera setups, and the acquired images were interpreted by biologists to catalog substrate and vegetation variables for each location. Available for each station are longitude and latitude, the dominant three benthic substrates, and vegetation type if applicable. While the categorization could be optimized to improve precision by using a continuous variable for substrate size, we have decided to preserve the domain-specific class-oriented format that the employees of the Government of Canada are used to in order to generate directly transferable results.

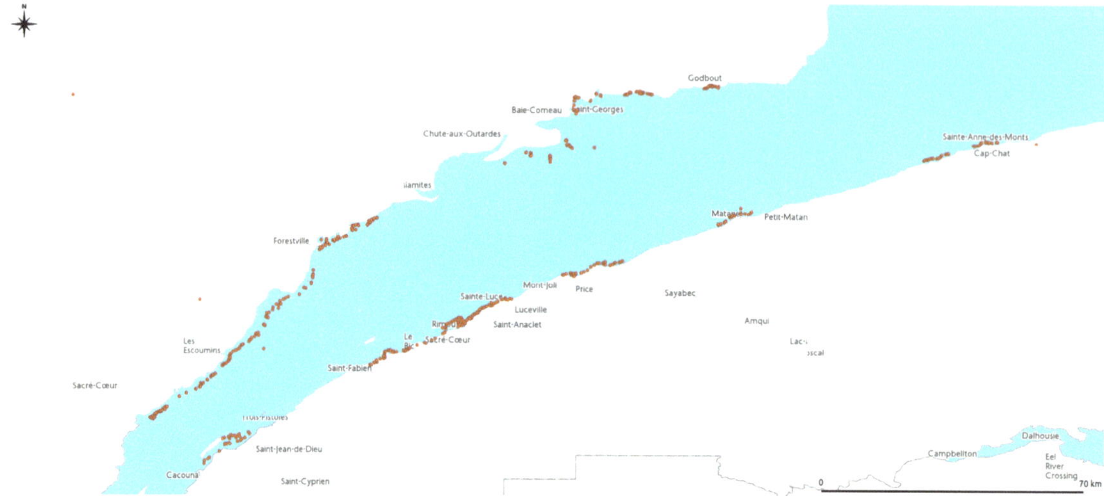

Figure 2. DFO ground-truth stations in the St. Lawrence River.

4.1.3. Training Data Generation

In order to generate the training data, the 905 ground-truth stations' data have been cross-referenced with the MBES data in order to obtain a neighborhood large enough to compute the feature vector for every ground station point. Ground-truth stations that did not fit the neighborhood requirements were discarded. Furthermore, to augment the data to consider multiple scales, soundings in a direct 3-meter neighborhood of the ground-truth stations were labeled with the class of their respective ground-truth station.

Habitat classes were derived from substrate particle size nomenclature provided by DFO in order to generate products meant to be helpful to biologists. Namely, we find bedrock, block, cobble, gravel, sand, and sandy mud classes.

4.2. Feature Engineering

The feature vector for each sounding is comprised of 16 geomorphometric variables (see Table 1). These variables represent a geometric signature for each sounding that allows us to numerically describe a sounding and its neighborhood. This selection of 16 features were chosen for their solid establishment into the literature on applied geomorphometry to the hydroghraphic field [16,17]. Given a sounding $p = (x, y, z)$, we consider a neighborhood N with $p_i \in N$, on which we compute a local covariance matrix with eigenvalues λ_1, λ_2, λ_3 and their respective eigenvectors e_1, e_2, e_3 :

Table 1. Feature space.

Feature	Recall
Sum	$\lambda_1 + \lambda_2 + \lambda_3$
Omnivariance	$(\lambda_1 \cdot \lambda_2 \cdot \lambda_3)^{1/3}$
Eigenentropy	$-\sum_{i=1}^{3} \lambda_i \cdot \ln \lambda_i$
Anisotropy	$(\lambda_1 - \lambda_2)/\lambda_3$
Planarity	$(\lambda_2 - \lambda_3)/\lambda_1$
Linearity	$(\lambda_1 - \lambda_2)/\lambda_1$
Surface variation	$\lambda_3/(\lambda_1 + \lambda_2 + \lambda_3)$
Sphericity	λ_3/λ_1
Verticality	$1 - \lvert(0,0,1) \cdot e_3\rvert$

Table 1. *Cont.*

Feature	Recall
First-Order Moment (w.r.t X axis)	$\sum_{i \in N}(p_i - p) \cdot e_1$
First-Order Moment (w.r.t Y axis)	$\sum_{i \in N}(p_i - p) \cdot e_2$
Second-Order Moment (w.r.t X axis)	$\sum_{i \in N}((p_i - p) \cdot e_1)^2$
Second-Order Moment (w.r.t Y axis)	$\sum_{i \in N}((p_i - p) \cdot e_2)^2$
Vertical Range	$z_{max} - z_{min}$
Height Above	$z - z_{min}$
Height Below	$z_{max} - z$

This gives us a 16-dimensional feature vector to work with.

4.3. Supervised Learning

Several supervised learning algorithms have been tested as part of the comparative analysis, whose results can be found in Table 2. The reference implementations were taken from the Scikit-Learn machine-learning library [20]. This choice was made to leverage existing work with an established governance structure, an open-source implementation, and a mature technological stack.

Table 2. Benchmarks (weighed averages over all classes).

Algorithm	Accuracy	Precision	Recall
K Nearest Neighbors	91.04%	90%	91%
Support Vector Machine	86.53%	77%	87%
Naive Bayes	73.60%	83%	74%
Adaboost	78.90%	85%	79%
Gradient Boosted Regression Trees	91.32%	91%	91%

4.3.1. K-Nearest Neighbors

First described by [21], the K-nearest neighbors algorithm (KNN) is based on the idea that points that share a high amount of similarities in their features also share similarities in the classification variable. As such, we can use the Euclidean distance over the 16-dimensional feature vector to create a useful similarity metric, under the implicit assumption that a shorter distance between two points implies a higher similarity between them.

4.3.2. Support Vector Machines (SVM)

Introduced by [22], the support vector machine has been a staple of both linear and non-linear classification problems due to its robustness. Anchored in the Vapnik–Chervonenkis statistical learning framework, the algorithm works on the principle of defining a boundary that maximizes the distance between categories by splitting the feature hyperspace using either hyperplanes in the linear case, or a kernel function in the non-linear case. Fresh data can then be categorized based on the class boundary.

4.3.3. Naive Bayes

The family of "naive" Bayes algorithms leverage Bayes' theorem under the assumption of independence between the feature variables, hence the "naive" qualifier. While this independence is not always grounded in reality, this can often be used as a weak signal for classification. As such, this is more of an industry-standard reference figure than an optimal classifier but can still yield interesting performances, especially when considering how simple the method is.

4.3.4. AdaBoost

Adaptive boosting, or Adaboost, is an ensemble method advanced in [23] that uses the output of several weak classifiers into a weighted sum to produce a strong classification. This meta-algorithm adapts the weights of each weak classifier's output signal to improve the classification of misclassified points. For optimal performance, the Adaboost-SAMAA variant with decision trees as weak multiclass classifiers is used [24].

4.3.5. Gradient Boosted Regression Trees

Gradient boosted regression trees is an ensemble method that uses successive approximations in order to generate a strong classification using a boosted set of weak decision tree classifiers. Building on the boosting theory of [23], the idea of combining boosting and stochastic gradient descent methods was introduced in [25] by using the residuals of each approximation step as the gradient of a loss-function to be minimized. Its claim to fame is that it generally performs better than random forests.

4.4. Unsupervised Learning

The fact that most of the world's seafloor substrates have not been sampled reveals a definite lack of ground-truthing data for supervised methods. In this context, new methods that do not rely on ground-truthing data are necessary to fill the gap. To this end, we have devised an unsupervised method based on a Gaussian mixture models (GMM) based on the same feature space to blindly classify substrates. The core concept of GMM methods is to classify data under the assumption that they are made of a finite number of independent Gaussian distributions. The separation between distributions is carried out by estimating the center of each Gaussian and its associate covariance structure [26]. In our case, we define the number of clusters as the number of substrates. To estimate the best fit in terms of cluster count, we use the Bayesian Information Criterion (BIC) [27], whose efficiency is well established in the literature with parameter estimation successes reaching 90% [28]. By finding the minimal BIC for an array of models, we thus find the best model fit for a given dataset. It is worth noticing that unlike supervised learning methods, this clustering is dependent on the underlying data and therefore does not generalize coherently to multiple independent datasets.

5. Results

By running models on each MBES dataset, we finally arrive at a fully classified dataset to generate habitat maps with GIS software. Here QGIS 3.22, a free and open-source GIS system, was used. The accuracy, precision, and recall of each supervised learning algorithm can be measured by using cross-validation with 20% of the dataset left out in the training phase to test out-of-bag soundings (Table 2).

5.1. Model Performance

Supervised model performances (Table 2) show that two algorithms stand out by their high accuracy rate, precision, and recall. The K nearest neighbors method comes in first with an accuracy of 91%, a precision of 90%, and a recall of 91%. Gradient boosted regression trees comes second with all three parameters to 91%. Since the gradient-boosted method gives the best performance, we shall use it from this point as the reference for supervised learning methods. Accuracy, precision and recall were computed as follows:

$$Accuracy = \frac{TruePositives + TrueNegatives}{TotalPositives + TotalNegatives} \tag{1}$$

$$Precision = \frac{TruePositives}{TruePositives + FalsePositives} \tag{2}$$

$$Recall = \frac{TruePositives}{TruePositives + FalseNegatives} \tag{3}$$

5.2. Comparative Analysis

Once trained, the supervised and unsupervised models have been applied on bathymetric datasets for a substrates classification in multiple areas of the North Shore of the St. Lawrence maritime estuary. Data created by the machine learning models are a sequence of positions (x,y,z) with an associated class with a numerical identifier. In both models, x,y, and z are the same values. In the case of the supervised model, the associated identifier is a number within the range of 0 to 5, respectively, corresponding to six different classes of substrates (block—0; cobble—1; gravel—2; bedrock—3; sand—4; and sandy mud—5). In the case of the unsupervised model, the identifier refers to distinct yet unknown classes, and the maximum number of different classes is determined by the number of clusters associated with the minimum BIC parameter. In order to qualitatively assess the comparison between the two methods, we analyze the correspondence between classes in each model.

To this end, a simple algorithm has been devised: a matrix mapping the classes of the supervised model as rows and the classes of the unsupervised model as columns is set up (Figure 3). The cell with the pair of supervised/unsupervised classes with the highest count is selected, and the rest of the cell's row and column are set to 0. The process is repeated until all corresponding pairs are found. This yields the following correspondence matrix for our data, which confirms that the unsupervised method follows the human-like intuition that classes that are geometrically different, from a geomorphometric point of view, do correspond to different substrates.

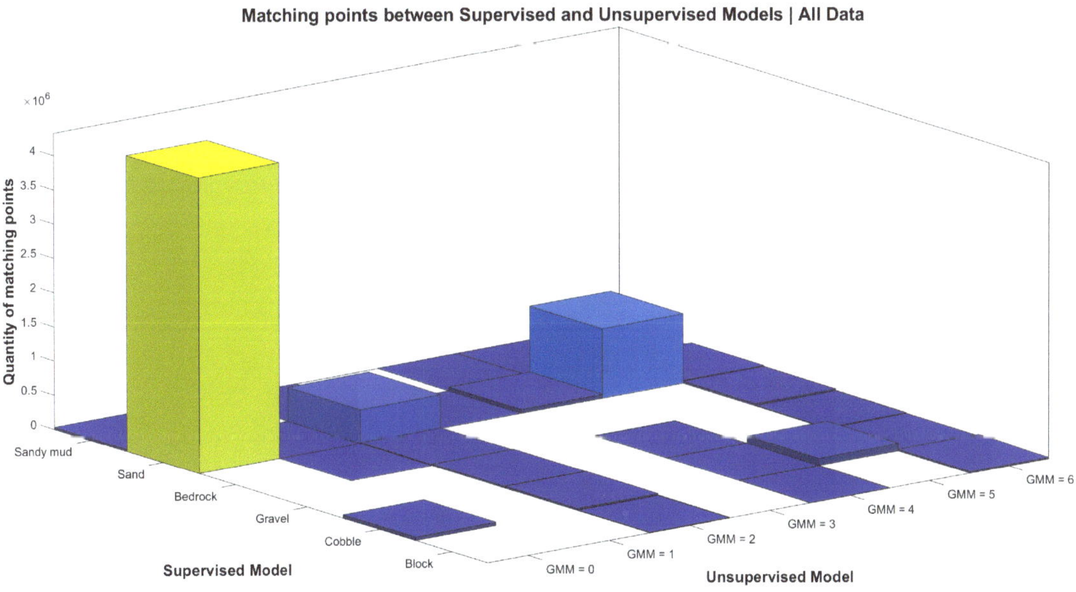

Figure 3. GBM and GMM correspondence.

5.3. Model-Generated Maps

Maps shown in this section (Figures 4–7) correspond to the supervised model data (a), the unsupervised model data (b), their matching area (c), and the 1-meter resolution bathymetry (d).

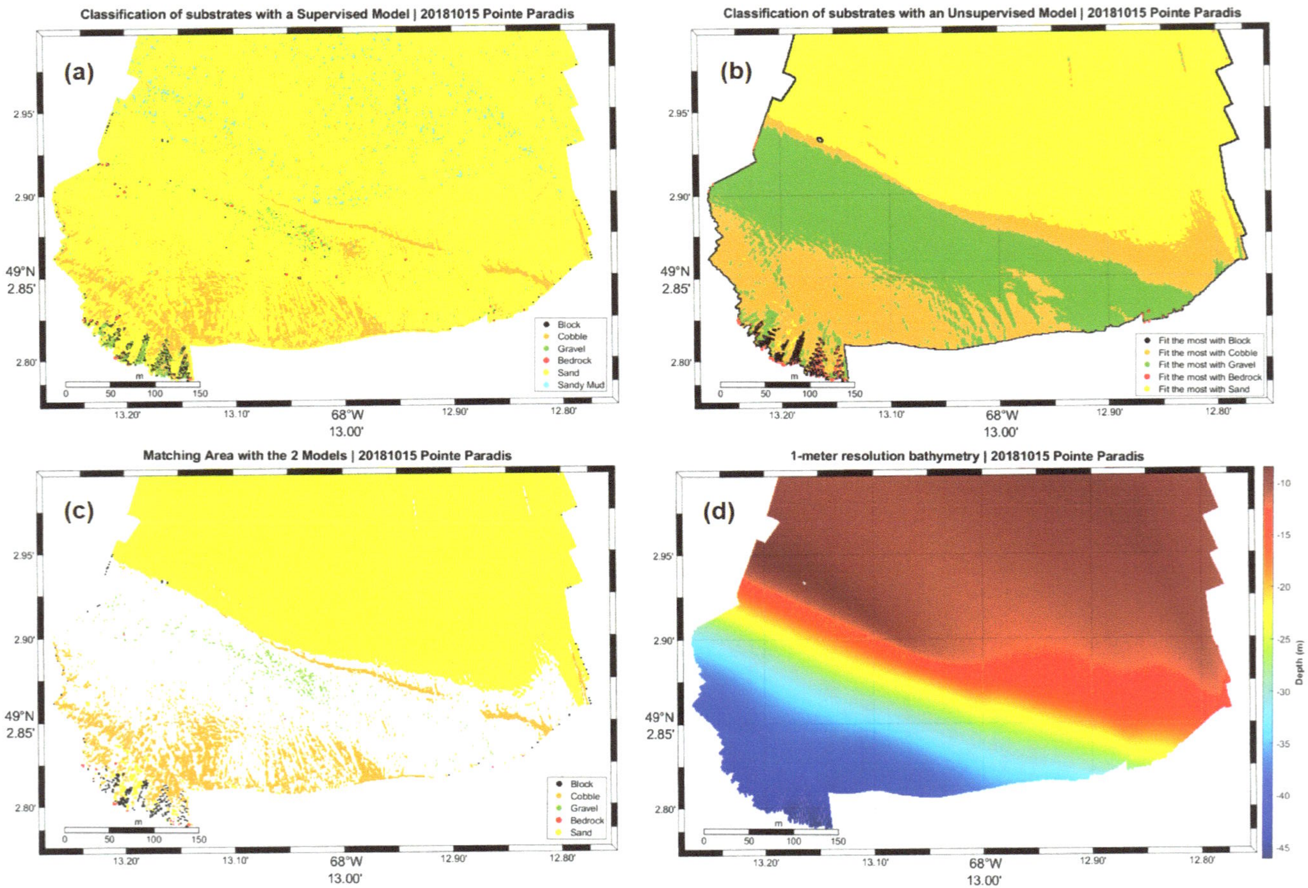

Figure 4. Pointe Paradis area with supervised model (**a**), unsupervised model (**b**), matching points between the two models (**c**) and the 1 m resolution bathymetry (**d**).

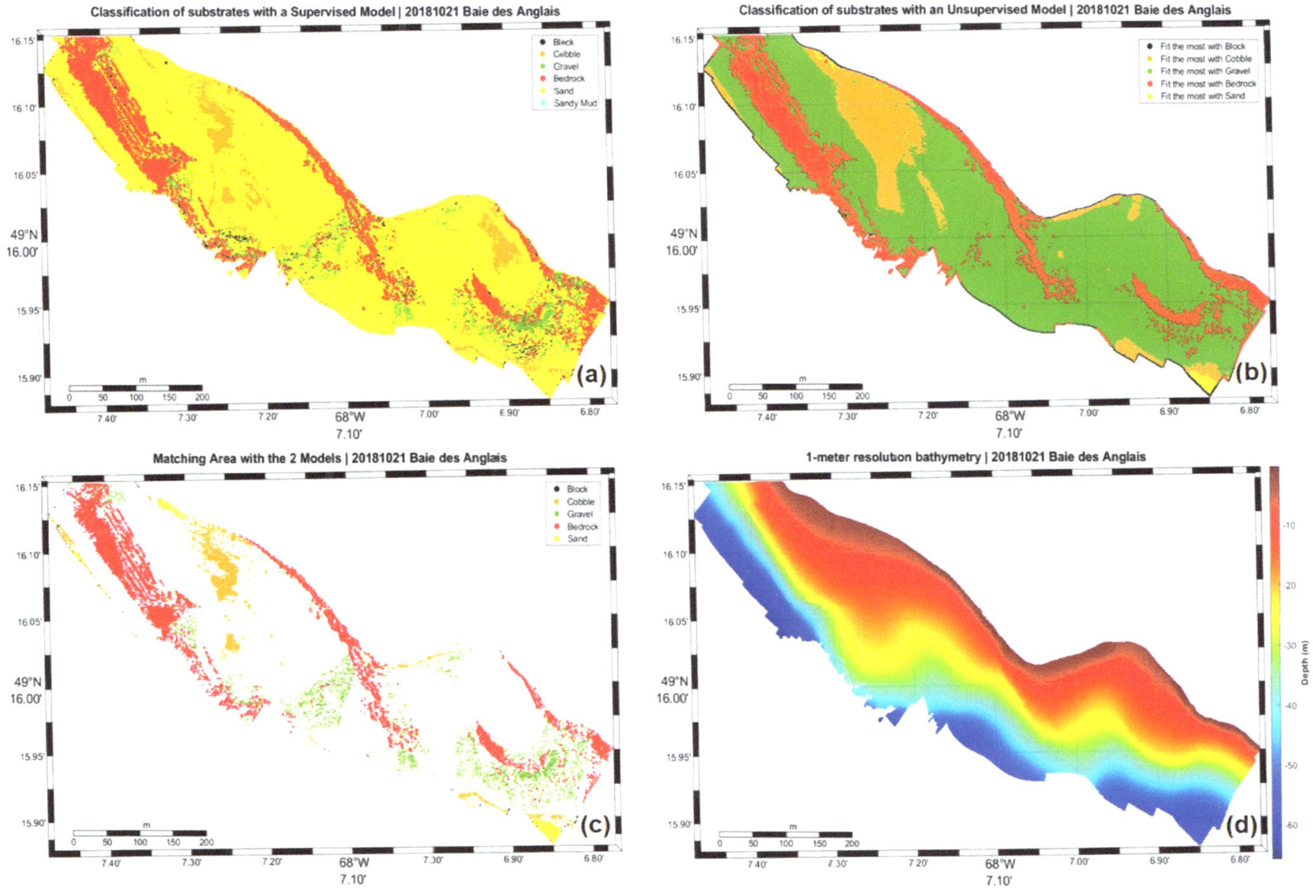

Figure 5. Baie des Anglais area with supervised model (**a**), unsupervised model (**b**), matching points between the two models (**c**) and 1 m resolution bathymetry (**d**).

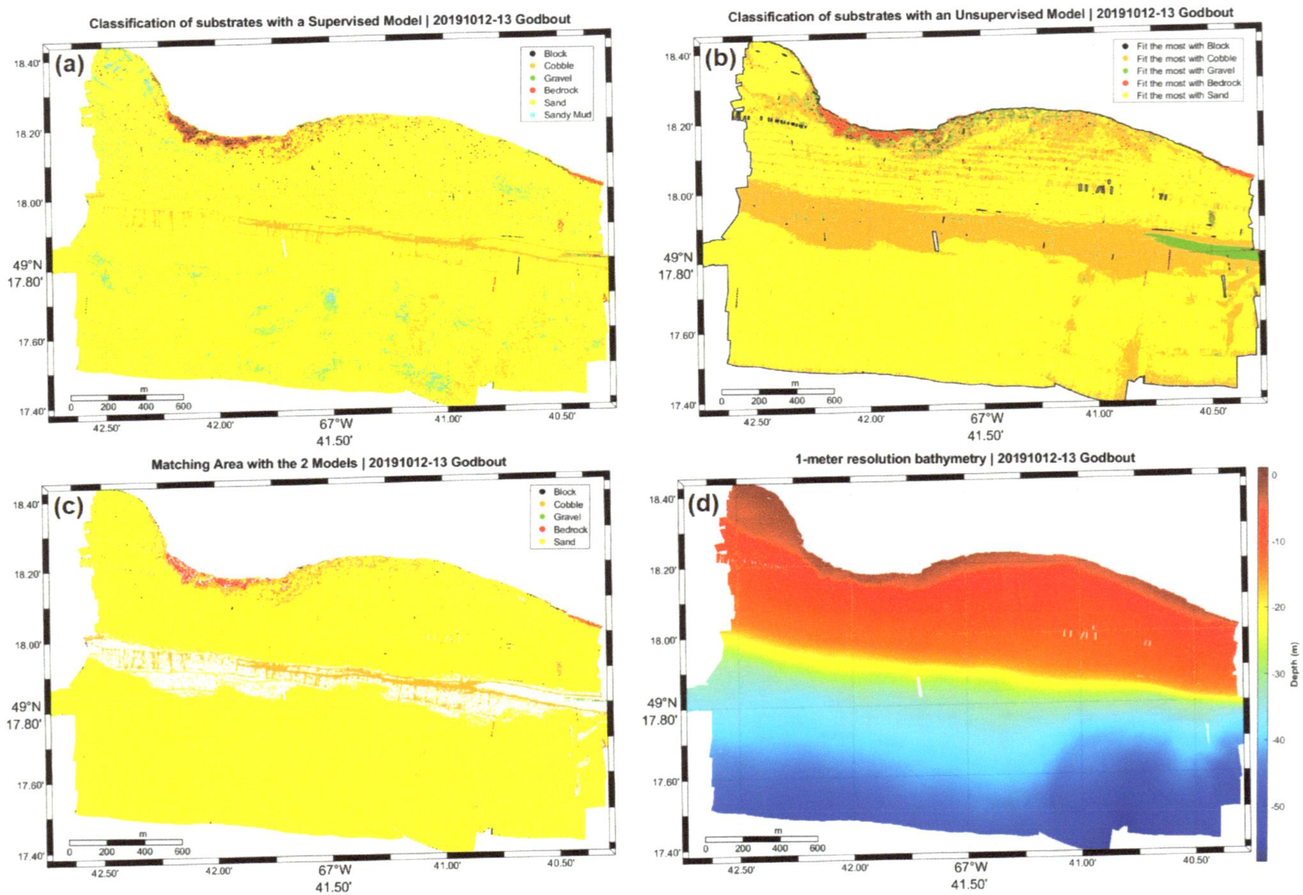

Figure 6. Godbout area with supervised model (**a**), unsupervised model (**b**), matching points between the two models (**c**) and the 1 m resolution bathymetry (**d**).

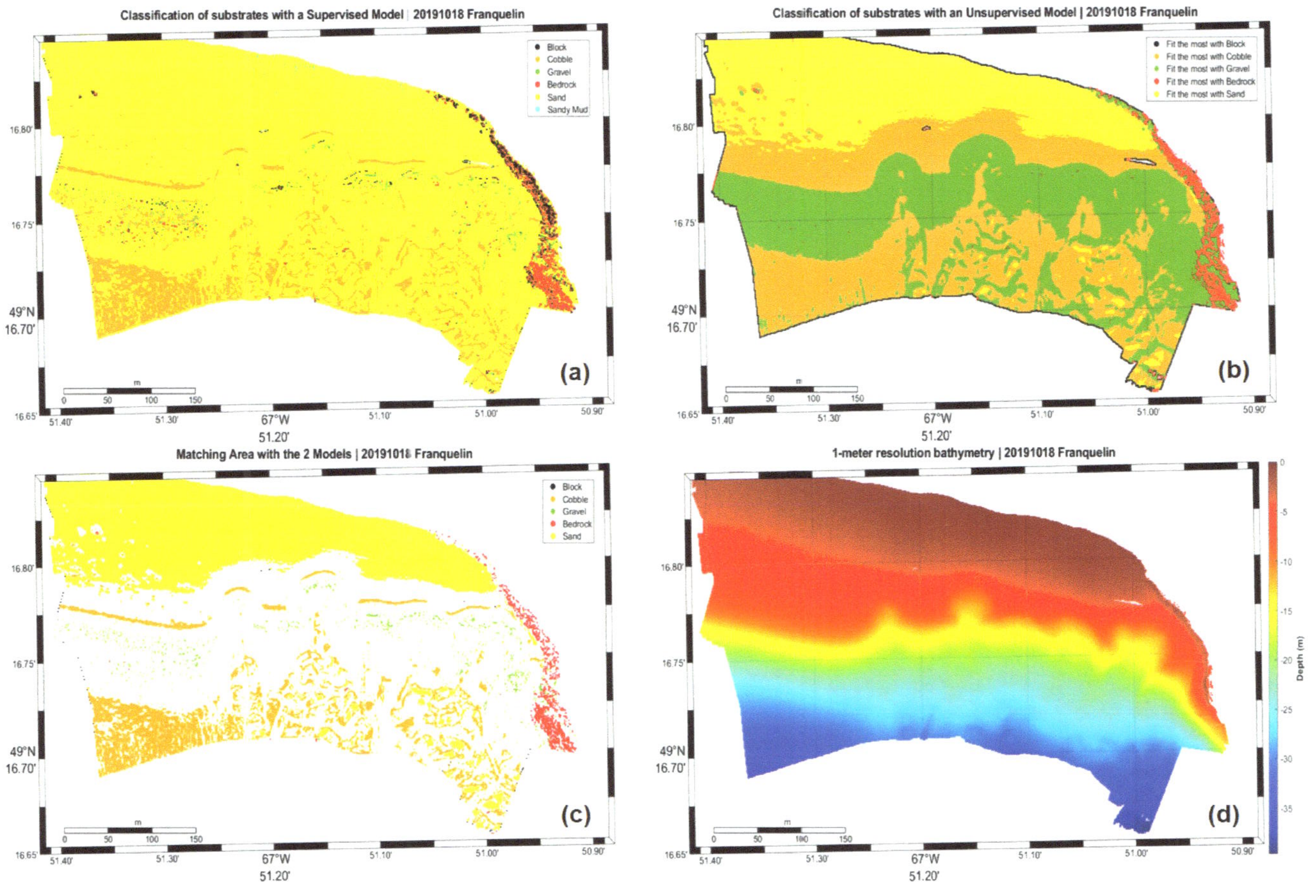

Figure 7. Franquelin area with supervised model (**a**), unsupervised model (**b**), matching points between the two models (**c**) and the 1 m resolution bathymetry (**d**).

6. Discussion

Substrate classification using supervised learning and unsupervised learning on geomorphometric features derived from bathymetric data allows us to create detailed maps of the North Shore of the St. Lawrence Estuary.

The design choice of using solely geomorphometric variables can be argued. Several other proxy variables could be leveraged to enhance the model. Of particular interest is the strength of the acoustic return from the seafloor recorded by the echosounder (backscatter), which has been shown to be a very effective predictor [29]. However, the acoustic strength is dependent on multiple environmental variables, and as such, lacks a common comparison basis. For example, the same seafloor may exhibit different acoustic return characteristics depending on the direction of surveying. Additionally, different sensors will yield different backscatter readings due to their inherent sensitivity differences. Research in the field of backscatter normalization should solve these issues in the future, but the current state of the art and the lack of availability of normalized backscatter datasets prevented it from being integrated into this research where a generalizable model was sought. Furthermore, relying solely on geomorphometric variables allows the method to be generalized to a large variety of remote sensing methods in addition to multibeam echosounders.

Supervised learning using gradient-boosting generates substrate classes based on ground-truth data (905 points distributed in all the maritime estuary). This model has a high performance based on its precision, accuracy, and recall (91%). On the other hand, even with the best performance, it is vulnerable to noise and artifacts, which can significantly alter the classification [30]. We have encountered such an event as shown in Figure 6 which is located in the Godbout area. Several artifacts emanating from calibration issues in the hydrographic system have confused the algorithm and led it to believe that a double line of cobbles was present instead of one at the 20-meter isobath. The proper equipment calibration is therefore critical before carrying out the hydrographic survey. The proper methodology must also be enforced to respect international standards with regard to hydrographic surveying best practices [31].

Additionally, supervised learning is highly dependent on the classification system used by domain experts. This methodology depends heavily on the quality [32] and quantity [33] of substrate samples, which are assumed to have been perfectly collected and tagged by field operatives who will expertly interpret their classification. It also can be argued that the geometric signatures derived using coastal data may not generalize well to offshore study zones and that further research is warranted to validate the claim that these signatures would be free of such bias. This allows for the introduction of human error and bias in the training data, which strongly highlights the need for establishing proper governance and procedures around data acquired for the purpose of training artificial bits of intelligence, both of which come at a significant cost that adds on top of the considerable cost of acquiring field data for marine sciences.

Unsupervised learning classifications based on Gaussian mixture models [26] do not require ground truthing data and can be applied to all areas where bathymetric data is available. It relies on an abstract geometrical distance definition, which has a nearly limitless range of possibilities in terms of model expansion to accommodate additional proxy variables. This flexibility and lack of a priori bias make it very suitable for exploratory analysis of the seafloor and anomaly detection. The Gaussian mixture model was chosen for ease of implementation and optimization through the minimum-BIC method. Further optimizations and model selection could be carried out as part of future research.

However, it follows from the arbitrary choice of substrate labels in the supervised learning model that not all classes have corresponding classes in each model, respectively. This is further exacerbated by the unsupervised model's geometric distance definition based on geomorphometric features, which do not distinguish the differential characteristics of each feature with regard to its substrate classification, which results in some features weighing more in the unsupervised model's perspective.

Furthermore, discrepancies between identical supervised classes referring to different unsupervised classes highlight the limitations of geomorphometric methods. Their dependence on geometric indicators can make them vulnerable to identical substrates forming different kinds of geometries, for example, sand formations that can be modeled into varied structures under the action of water currents, yielding different kinds of patterns such as dunes and such. This could be a promising area for further research.

Additionally, both models exhibit non-negligible edge effects at the extremities of the map, where the eigenvalues of the point's neighborhood's covariance matrix is either unstable of singular. This effectively results in the creations of singularities like in Figures 4b, 5b, 6b and 7b.

Thus, the first results created with the unsupervised model are promising. Eventually, the current version of this model could be useful for flat sand-dominated areas and, to a lesser extent, for bedrock-dominated areas in an environmental study setting.

Another criticism of the method is the fact that both models are only trained on a 1-meter resolution bathymetry grid. Depending on the resolution of the seafloor required, both models could benefit from performing a multi-resolution analysis to adequately capture macroscopic phenomenons at different scales. This would effectively increase the number of parameters by multiplying the current parameter count by the number of different scales. These new models could be used to identify patterns that could not have been seen with a one-meter resolution dataset [34].

7. Conclusions

The use of machine learning techniques to predict substrate classes based on geomorphometric features allows for promising habitat modeling processes. As such, the technique has proven itself to be very useful to efficiently map out benthic habitats and drastically reduce costs. As such, the techniques developed here can be leveraged as powerful automation mechanisms to open doors in improving resource monitoring in science, industry, and everywhere information on the characteristics of the seafloor is useful.

The use of a sparse DTM implies that the technique can be readily generalized to data coming from a large variety of sensors such as multibeam echosounders, aerial bathymetric lidar, satellite-derived bathymetry, and many more. Further research could easily leverage more than one remote sensing method to improve on this technique.

8. Open Source Software

The software developed in this research project is made available under MIT license at the following address: https://github.com/CIDCO-dev/BenthicClassifier (accessed on 28 June 2024).

Author Contributions: Conceptualization, G.L.-M.; methodology, G.L.-M. and T.L.; software, G.L.-M., T.L., P.C.-M., D.G. and D.D.; validation, G.L.-M. and T.L.; formal analysis, G.L.-M., T.L., P.C.-M. and D.G.; investigation, G.L.-M., T.L., P.C.-M., D.G., D.D., M.-A.C. and D.N.M.; resources, D.D., M.-A.C. and D.N.M.; data curation, M.-A.C. and D.N.M.; writing—original draft preparation, G.L.-M. and T.L.; writing—review and editing, G.L.-M. and T.L.; visualization, T.L., P.C.-M. and D.G.; supervision, G.L.-M.; project administration, G.L.-M.; funding acquisition, G.L.-M. All authors have read and agreed to the published version of the manuscript.

Funding: This research was made possible with the support of the Department of Fisheries and Oceans from the Government of Canada under the Coastal Environmental Baseline Program.

Data Availability Statement: The raw data supporting the conclusions of this article will be made available by the authors on request.

Conflicts of Interest: The authors declare no conflicts of interest. The funders had no role in the design of the study; in the collection, analyses, or interpretation of data; in the writing of the manuscript; or in the decision to publish the results.

Abbreviations

The following abbreviations are used in this manuscript:

BIC	Bayesian Information Criterion
DFO	Department of Fisheries and Oceans Canada
DTM	Digital Terrain Model
GBM	Gradient Boosting Method
GIS	Geographic Information System
GMM	Gaussian Mixture Method
GNSS	Global Navigation Satellite System
INS	Inertial Navigation System
MBES	Multibeam Echosounder
OECD	Organisation for Economic Co-operation and Development

References

1. NOAA. How Much of the Ocean Have We Explored? National Ocean Service Website. 2023. Available online: https://oceanexplorer.noaa.gov/facts/explored.html (accessed on 15 May 2023).
2. Proudfoot, B.; Devillers, R.; Brown, C.J.; Edinger, E.; Copeland, A. Seafloor mapping to support conservation planning in an ecologically unique fjord in Newfoundland and Labrador, Canada. *J. Coast. Conserv.* **2020**, *24*, 36. [CrossRef]
3. Flannagan, J.F. Efficiencies of Various Grabs and Corers in Sampling Freshwater Benthos. *J. Fish. Res. Board Can.* **1970**, *27*, 1691–1700. [CrossRef]
4. Bouma, A. *Methods for the Study of Sedimentary Structures*; Wiley-Interscience: New York, NY, USA, 1969.
5. Snellen, M.; Gaida, T.C.; Koop, L.; Alevizos, E.; Simons, D.G. Performance of multibeam echosounder backscatter-based classification for monitoring sediment distributions using multitemporal large-scale ocean data sets. *IEEE J. Ocean. Eng.* **2018**, *44*, 142–155. [CrossRef]
6. Lurton, X.; Pouliquen, E. Automated Sea-bed Classification System For Echo-Sounders. In Proceedings of the OCEANS 92 Proceedings@m_Mastering the Oceans Through Technology, Newport, RI, USA, 26–29 October 1992; Volume 1, pp. 317–321.
7. Pillay, T.; Cawthra, H.; Lombard, A. Characterisation of seafloor substrate using advanced processing of multibeam bathymetry, backscatter, and sidescan sonar in Table Bay, South Africa. *Mar. Geol.* **2020**, *429*, 106332. [CrossRef]
8. Horning, N.; Robinson, J.A.; Sterling, E.J.; Turner, W. *Remote Sensing for Ecology and Conservation: A Handbook of Techniques*; Oxford University Press: Oxford, UK, 2010.
9. Lyzenga, D.R.; Malinas, N.P.; Tanis, F.J. Multispectral bathymetry using a simple physically based algorithm. *IEEE Trans. Geosci. Remote Sens.* **2006**, *44*, 2251–2259. [CrossRef]
10. Ma, S.; Tao, Z.; Yang, X.; Yu, Y.; Zhou, X.; Li, Z. Bathymetry Retrieval From Hyperspectral Remote Sensing Data in Optical-Shallow Water. *IEEE Trans. Geosci. Remote Sens.* **2014**, *52*, 1205–1212. [CrossRef]
11. Haralick, R.M.; Shanmugam, K.; Dinstein, I.H. Textural features for image classification. *IEEE Trans. Syst. Man Cybern.* **1973**, *SMC-3*, 610–621. [CrossRef]
12. Geisz, J.K.; Wernette, P.A.; Esselman, P.C. Classification of Lakebed Geologic Substrate in Autonomously Collected Benthic Imagery Using Machine Learning. *Remote Sens.* **2024**, *16*, 1264. [CrossRef]
13. Wahidin, N.; Siregar, V.P.; Nababan, B.; Jaya, I.; Wouthuyzen, S. Object-based Image Analysis for Coral Reef Benthic Habitat Mapping with Several Classification Algorithms. *Procedia Environ. Sci.* **2015**, *24*, 222–227. [CrossRef]
14. Jackett, C.; Althaus, F.; Maguire, K.; Farazi, M.; Scoulding, B.; Untiedt, C.; Ryan, T.; Shanks, P.; Brodie, P.; Williams, A. A benthic substrate classification method for seabed images using deep learning: Application to management of deep-sea coral reefs. *J. Appl. Ecol.* **2023**, *60*, 1254–1273. [CrossRef]
15. Arosio, R.; Hobley, B.; Wheeler, A.J.; Sacchetti, F.; Conti, L.A.; Furey, T.; Lim, A. Fully convolutional neural networks applied to large-scale marine morphology mapping. *Front. Mar. Sci.* **2023**, *10*, 1228867. [CrossRef]
16. Hackel, T.; Wegner, J.D.; Schindler, K. Fast Semantic Segmentation of 3D Point Clouds With Strongly Varying Density. *Isprs Ann. Photogramm. Remote Sens. Spat. Inf. Sci.* **2016**, *3*, 177–184. [CrossRef]
17. Lecours, V.; Dolan, M.F.; Micallef, A.; Lucieer, V.L. A review of marine geomorphometry, the quantitative study of the seafloor. *Hydrol. Earth Syst. Sci.* **2016**, *20*, 3207–3244. [CrossRef]
18. Misiuk, B.; Brown, C.J. Benthic habitat mapping: A review of three decades of mapping biological patterns on the seafloor. *Estuar. Coast. Shelf Sci.* **2023**, *296*, 108599. [CrossRef]
19. Calder, B. Automatic Statistical Processing of Multibeam Echosounder Data. *Int. Hydrogr. Rev.* **2003**, *4*, 53–68.
20. Pedregosa, F.; Varoquaux, G.; Gramfort, A.; Michel, V.; Thirion, B.; Grisel, O.; Blondel, M.; Prettenhofer, P.; Weiss, R.; Dubourg, V.; et al. Scikit-learn: Machine Learning in Python. *J. Mach. Learn. Res.* **2011**, *12*, 2825–2830.
21. Fix, E.; Hodges, J.L. Discriminatory Analysis. Nonparametric Discrimination: Consistency Properties. *Int. Stat. Rev./Rev. Int. De Stat.* **1989**, *57*, 238–247. [CrossRef]

22. Boser, B.E.; Guyon, I.M.; Vapnik, V.N. A training algorithm for optimal margin classifiers. In *COLT '92: Proceedings of the Fifth Annual Workshop on Computational Learning Theory, Pittsburgh, PA, USA, 27–29 July 1992*; ACM Press: New York, NY, USA, 1992; pp. 144–152.
23. Freund, Y.; Schapire, R.E. A Short Introduction to Boosting. In Proceedings of the Sixteenth International Joint Conference on Artificial Intelligence, Stockholm, Sweden, 31 July–6 August 1999; Morgan Kaufmann: Burlington, MA, USA, 1999; pp. 1401–1406.
24. Hastie, T.; Rosset, S.; Zhu, J.; Zou, H. Multi-class AdaBoost. *Stat. Its Interface* **2009**, *2*, 349–360. [CrossRef]
25. Friedman, J.H. Stochastic Gradient Boosting. *Comput. Stat. Data Anal.* **2002**, *38*, 367–378. [CrossRef]
26. Reynolds, D.A. Gaussian mixture models. *Encycl. Biom.* **2009**, *741*, 659–663.
27. Schwarz, G. Estimating the dimension of a model. *Ann. Stat.* **1978**, 461–464. [CrossRef]
28. Steele, R.; Raftery, A.E. Performance of Bayesian model selection criteria for Gaussian mixture models. *Front. Stat. Decis. Mak. Bayesian Anal.* **2010**, *2*, 113–130.
29. Brown, C.J.; Beaudoin, J.; Brissette, M.; Gazzola, V. Multispectral Multibeam Echo Sounder Backscatter as a Tool for Improved Seafloor Characterization. *Geosciences* **2019**, *9*, 126. [CrossRef]
30. Gupta, S.; Gupta, A. Dealing with noise problem in machine learning data-sets: A systematic review. *Procedia Comput. Sci.* **2019**, *161*, 466–474. [CrossRef]
31. Mills, G.B. International hydrographic survey standards. *Int. Hydrogr. Rev.* **1998**. [CrossRef]
32. Long, D. BGS Detailed Explanation of Seabed Sediment Modified Folk Classification. MESH (Mapping European Seabed Habitats). 2006. Available online: https://webarchive.nationalarchives.gov.uk/ukgwa/20101014090013/http://www.searchmesh.net/PDF/GMHM3_Detailed_explanation_of_seabed_sediment_classification.pdf (accessed on 15 May 2023).
33. Weiss, G.M.; Provost, F. Learning when training data are costly: The effect of class distribution on tree induction. *J. Artif. Intell. Res.* **2003**, *19*, 315–354. [CrossRef]
34. Blayvas, I.; Kimmel, R. Machine learning via multiresolution approximation. *Ieice Trans. Inf. Syst.* **2003**, *86*, 1172–1180.

MDPI AG
Grosspeteranlage 5
4052 Basel
Switzerland
Tel.: +41 61 683 77 34

Geomatics Editorial Office
E-mail: geomatics@mdpi.com
www.mdpi.com/journal/geomatics

www.ingramcontent.com/pod-product-compliance
Lightning Source LLC
Chambersburg PA
CBHW041347210926
51850CB00013B/3018